Lecture Notes in Mathematics 1532

Editors:
A. Dold, Heidelberg
B. Eckmann, Zürich
F. Takens, Groningen

T0222870

Jean François Colombeau

Multiplication of Distributions

A tool in mathematics, numerical engineering
and theoretical physics

Springer-Verlag

Berlin Heidelberg New York
London Paris Tokyo
Hong Kong Barcelona
Budapest

Author

Jean François Colombeau
Ecole Normale Supérieure de Lyon
46 Allée d'Italie
F-69364 Lyon Cedex 07, France

Mathematics Subject Classification (1991): 03H05, 26E35, 30G99, 35A40, 35D05, 35L60, 35R05, 46F10, 65M05, 73D05, 76L05, 76T05

ISBN 3-540-56288-5 Springer-Verlag Berlin Heidelberg New York
ISBN 0-387-56288-5 Springer-Verlag New York Berlin Heidelberg

© Springer-Verlag Berlin Heidelberg 1992
Printed in Germany

Typesetting: Camera ready by author
Printing and binding: Druckhaus Beltz, Hemsbach/Bergstr.
46/3140-543210 - Printed on acid-free paper

Introduction

The aim of this book is to present a recent mathematical tool, in a way which is very accessible and free from mathematical techniques. The presentation developed here is in part heuristic, with emphasis on algebraic calculations and numerical recipes that can be easily used for numerical solutions of systems of equations modelling elasticity, elastoplasticity, hydrodynamics, acoustic diffusion, multifluid flows. This mathematical tool has also theoretical consequences such as convergence proofs for numerical schemes, existence - uniqueness theorems for solutions of systems of partial differential equations, unification of various methods for defining multiplications of distributions. These topics are not developed in this book since this would have made it not so elementary. A glimpse on these topics is given in two recent research expository papers : Colombeau [14] in Bull. of A.M.S. and Egorov [1] in Russian Math. Surveys. A detailed and careful self contained exposition on these mathematical applications can be found in Oberguggenberger's recent book [11] " Multiplication of distributions and applications to partial differential equations". A set of references is given concerning both the applied and the theoretical viewpoints. This book is the text of a course in numerical modelling given by the author to graduate students at the Ecole Normale Supérieure de Lyon in the academic years 1989 - 90 and 1990 - 91.

Many basic equations of physics contain, in more or less obvious or hidden ways, products looking like "ambiguous multiplications of distributions" such as products of a discontinuous function f and a Dirac mass centered on a point of discontinuity of f or powers of a Dirac mass. These products do not make sense within classical mathematics (i. e. distribution theory) and usually appear as "ambiguous" when considered from a heuristic or physical viewpoint. The idea developed here is that these statements of equations of physics are basically sound, and that a new mathematical theory of generalized functions is needed to explain and master them. Such a theory was first developed in pure mathematics and then it was used in applications ; the mathematician reader can look at the books Colombeau [2, 3], Part II of Rosinger [1], Biagioni [1] and Oberguggenberger [11].

The ambiguity appearing in equations of physics when these equations involve "heuristic multiplications of distributions" corresponds in our theory to the fact that, when formulated in the weakest way, these equations have an infinite number of possible solutions. This recognition of infinitely many solutions was essentially known and understood without our theory (at least in Quantum Field Theory). To resolve the ambiguities our new setting can suggest more precise formulations of the equations (these more precise formulations do not make sense within distribution theory). On physical ground one chooses one of these more precise formulations in which there is no more ambiguity. This technique is developed in this book on various examples from physics. This gives directly new algebraic formulas and new numerical schemes. When one has algebraic jump formulas (for systems in nonconservation form) then it is an easy further step to transfer this knowledge into numerical schemes of the Godunov type. This last numerical technique - Godunov schemes for systems in nonconservation form (elastoplasticity, multifluid flows) or for nonconservative versions of systems of conservation laws (hydrodynamics) - is the main application developed in this text (chapters 4 and 5).

The book is divided into four parts. Part I (chapters 1 and 2) deals with preliminaries from mathematics and physics. Part II (chapter 3) is a smooth introduction to our theory of generalized functions. Part III (chapters 4, 5, 6) is the main part : there new numerical methods are developed ; for simplicity most of them are presented on one dimensional models, but they extend to the 2 and 3 dimensional problems of industrial use or physical significance ; numerical results are presented and references are given. Part IV is made of various complements.

Now let us describe briefly the contents of each chapter. In chapter 1 we introduce our viewpoint, distribution theory and its limitations, in a way convenient for a reader only aware of the concepts of partial derivatives (of functions of several real variables) and of integrals (of continuous functions). Chapter 2 exposes the main equations of Continuum Mechanics considered in the book (hydrodynamics, elastoplasticity, multifluid flows, linear acoustics). The aim of chapter 3 is to describe this new mathematical tool without giving the precise mathematical definitions : the viewpoint there is that these generalized functions can be manipulated correctly provided one has an intuitive understanding of them and provided one is familiar with their rules of calculation. Chapter 4 deals with the classical (conservative) system of fluid dynamics. No products of distributions appear in it , even in case of shock waves. But, surprisingly, our tool gives new methods for its numerical solution : one transforms it into a simpler, but in nonconservative form, system and then one computes a solution from nonconservative Godunov type schemes. In this case, since the correct solution is known with arbitrary precision it is easy to evaluate the value of the new method (by comparison with the exact solution and with numerical results from classical conservative numerical methods). Chapters 5 and 6 deal with systems containing multiplications of distributions that arise directly from physics : nonlinear systems of elastoplasticity and multifluid flows in chapter 5 and linear systems of acoustics in chapter 6. In chapter 7 we expose in the case of a simple model (a self interacting boson field) the basic heuristic calculations of Quantum Field Theory. This topic has been chosen since Quantum Field Theory is the most famous historic example in which the importance of multiplications of distributions was first recognized. Chapter 8 contains a mathematical introduction to these generalized functions and mathematical definitions.

I am particularly indebted to A. Y. Le Roux and B. Poirée. I was working on the multiplication of distributions from a viewpoint of pure mathematics when we met . Their research work (numerical analysis and engineering, physics) had shown them the need of a multiplication of distributions. They introduced me kindly and smoothly to their problems. This was the origin of the present book. I am also very much indebted to L. Arnaud, F. Berger, H. A. Biagioni, L.S. Chadli, P. De Luca, J. Laurens, A. Noussaïr, M. Oberguggenberger, B. Perrot, I. Zalzali for help in works used in the preparation of this book. The main part of the typing has been done by B. Mauduit to whom I also extend my warmest thanks.

Table of contents

Part I. The mathematical and physical contexts

Chapter 1. Introduction to generalized functions and distributions.

Chapter 2. Multiplications of distributions in classical physics

Part II. Generalized Functions

Chapter 3. Elementary introduction.

Part III. Algebraic and Numerical Solutions of Systems of Continuum Mechanics : Hydrodynamics, Elasticity, Elastoplasticity, Acoustics.

Chapter 4. Jump formulas for systems in nonconservation form. New numerical methods.

Chapter 1.   Introduction to generalized functions and distributions

§1.1 THE VIEWPOINT OF THIS BOOK.

Long ago physicists and engineers have introduced formal calculations that work well (Heaviside [1] , Dirac [1]) ; in particular they have introduced the Dirac delta function on $\mathbb{R}$

$$\delta(x) = 0 \quad \text{if} \quad x \neq 0$$

$$\delta(0) = +\infty \text{ (so "large" that } \int_{-\infty}^{+\infty} \delta(x) \, dx = 1).$$

Intuitively the Dirac delta function can be considered as some kind of limit - in a sense to be made precise-of the functions δ^ε when $\varepsilon \to 0$: support of $\delta^\varepsilon \subset [-\eta(\varepsilon), \eta(\varepsilon)]$ with $\eta(\varepsilon) \to 0$ as $\varepsilon \to 0$, $\int \delta^\varepsilon(x) \, dx = 1).$

The theory of distributions (Sobolev [1], Schwartz [1]) has given a rigorous mathematical sense to δ and other objects. But, in some important cases, in which the calculations of physicists are complicated (and give results in agreement with those from experiments at the price of ad hoc manipulations : for instance renormalization theory in Quantum Field Theory), the theory of distributions fails. One observes that these formal calculations involve unjustified products of distributions, such as δ^2, the square of the Dirac delta function. L. Schwartz [2] has proved in 1954 the "impossibility of the multiplication of distributions", even in a framework rather disjoint from the theory of distributions. He has proved the non-existence of a differential algebra A (of any kind of "generalized functions" on $\mathbb{R}$) containing the algebra $\mathcal{C}(\mathbb{R})$ (of continuous functions on $\mathbb{R}$) as a subalgebra, preserving the differentiation of functions of class $\mathcal{C}^1$ (i. e. the differentiation in A coincides with the classical one) and having a few other natural properties (Leibniz's rule for the differentiation of a product, the constant function 1 is the neutral element in A for the multiplication, A contains some version of the Dirac delta function). Thus the theory of distributions is not really concerned with this impossiblility result and it appears that the roots of the impossibility go back as far as some incoherence between the multiplication and the differentiation in the setting of $\mathcal{C}^1$ and continuous functions. Since the emergence and recent development of computer science other kinds of "multiplications of distributions " are successfully treated numerically, to model various problems from Continuum Mechanics (elasticity, elastoplasticity, hydrodynamics, acoustics, electromagnetism, ... see the sequel of this book).Therefore, there is presently a situation of impasse between (rigorous) mathematics from one side, theoretical physics and engineering from the other side.

Even, from a viewpoint internal to mathematics, one also faces a major problem : for most systems of partial differential equations (including those modelling the more usual physical situations) distribution solutions are unknown ; even, in many cases, one can prove - often trivially - the nonexistence of distribution solutions. This motivates the introduction of new mathematical objects (for instance the concept of "measure valued solutions to systems of conservation laws", see Di Perna [1], Di Perna - Majda [1], ...). This need, although internal to mathematics, is indeed closely related to the needs from physics and engineering (described above) since these explicit calculations or these numerical recipes are nothing other than attempts for the resolution of equations.

We shall present a mathematical theory of generalized functions, in which the main calculations and numerical recipes alluded to above make sense. This theory has recently been used in Continuum Mechanics, for problems involving "multiplications of distributions". It permits to understand the nature of the problems involved in these multiplications and it leads to new algebraic formulas (jump conditions for systems in nonconservative form), and new numerical methods. From a more theoretical viewpoint this theory gives solutions for previously unsolvable equations. In physically relevant cases these solutions can be indeed classical discontinuous functions (representing shock waves) which are not solutions within distribution theory. Since this theory is recent its limitations are still unknown and we propose numerous research directions (applied and theoretical).

If Ω denotes any open set in $\mathbb{R}^n$ we shall define a new concept of generalized functions on Ω (real valued or complex valued, even vector valued if needed ; we consider that they are real valued, unless the converse is explicitly stated). The set of these generalized functions is denoted by $\mathcal{G}(\Omega)$; $\mathcal{G}(\Omega)$ is a differential algebra (i. e. it has the same operations and rules as the familiar differential algebra $\mathcal{C}^{\infty}(\Omega)$ of all $\mathcal{C}^{\infty}$ functions on Ω). If $\mathcal{D}'(\Omega)$ is the vector space of all distributions on Ω (whose definition will be recalled in the next section) one has the inclusions

$$\mathcal{C}^{\infty}(\Omega) \subset \mathcal{D}'(\Omega) \subset \mathcal{G}(\Omega).$$

$\mathcal{G}(\Omega)$ induces on $\mathcal{D}'(\Omega)$ its addition, scalar multiplication and differentiation (there is no general multiplication in $\mathcal{D}'(\Omega)$). $\mathcal{G}(\Omega)$ induces on $\mathcal{C}^{\infty}(\Omega)$ all the operations in $\mathcal{C}^{\infty}(\Omega)$, in particular the multiplication. Thus these generalized functions look as some "super concept of $\mathcal{C}^{\infty}$ functions". The product of two arbitrary elements of $\mathcal{D}'(\Omega)$ will be in $\mathcal{G}(\Omega)$, not in $\mathcal{D}'(\Omega)$ in general. The connection with the Schwartz impossibility result is - as this could be expected - rather subtle. The algebra $\mathcal{C}(\Omega)$ of all continuous functions on Ω is not a subalgebra of $\mathcal{G}(\Omega)$: the product in $\mathcal{G}(\Omega)$ of two arbitrary continuous functions on Ω does not always coincide with their classical product. The subtility lies in that the difference (in $\mathcal{G}(\Omega)$) between these two products is - in some sense to be made precise after $\mathcal{G}(\Omega)$ will be defined - "infinitesimal" (although nonzero). Being "infinitesimal" this difference can be considered as null as long as it is not multiplied by some "infinite quantity" (infinite quantities like the value δ (0) of the Dirac delta function at the origin make sense in our setting). In all classical calculations dealing with continuous functions there do not appear such "infinite quantities" and so the difference between the two products of continuous functions is then always insignificant. The new theory is totally coherent with classical analysis and, at the same time, it escapes from Schwartz impossibility result.

The above should not sound too much mysterious, since physicists and mathematicians are indeed familiar with certain aspects of this subtility. Let us consider the following classical remark from shock wave solutions of systems of conservation laws (Richtmyer [1]).Consider the equation

(1) $\quad u_t + u u_x = 0$

and seek a travelling wave solution : i. e. $u(x, t) = a$ for $x < ct$, $u(x, t) = a+b$ for $x > ct$, a, b, constants, c is the constant velocity of the shock. Such a solution can be written as

(2) $\quad u(x, t) = b\, Y(x - ct) + a$

where Y is the Heaviside step function $(Y(\xi) = 0$ if $\xi < 0$, $Y(\xi) = 1$ if $\xi > 0)$. Interpreting (1) as

$$u_t + \frac{1}{2}(u^2)_x = 0$$

(2) gives (Y' is the derivative of Y in the sense of distributions, see next section)

$$- bc\, Y'(x - ct) + \frac{1}{2}(b^2\, Y(x - ct) + 2ab\, Y(x - ct) + a^2)_x = 0$$

since one has $Y^2 = Y$ (in the algebra of piecewise constant functions). One obtains (since Y' is non zero)

(3) $\quad c = a + \dfrac{b}{2}$.

Now let us multiply (1) by u : this gives

(4) $\quad u u_t + u^2 u_x = 0$

that can be naturally interpreted as

(4') $\quad \frac{1}{2}(u^2)_t + \frac{1}{3}(u^3)_x = 0$.

Putting (2) into (4) one gets for c a value different from (3). One concludes that (1) and (4) have different shock wave solutions : thus the correct statement of the equations has to be carefully selected on physical ground, see Richtmyer [1]. The passage from (1) to (4) is a multiplication, and so we have put in evidence some incoherence between multiplication and differentiation. This incoherence is reproduced in the following calculations. Classically one has

(5) $Y^n = Y$ $\forall n = 2, 3, ...$

Differentiation of (5) gives

(6) $n\, Y^{n-1}\, Y' = Y'$

thus one has

(7) $2Y\, Y' = Y'.$

Multiplication by Y gives

$$2Y^2 Y' = YY'.$$
Using (6) one gets

$$\frac{2}{3} Y' = \frac{1}{2} Y'$$

which is absurd since $Y' \neq 0$. Of course the trouble arises at the origin, since this is the unique singular point of Y and Y'. If one accepts to consider $Y^n \neq Y$ (n = 2, 3, ...) there is no more trouble. Of course $Y^n - Y$ is "infinitesimal" - in a sense to be made precise later, so that if- instead of multiplying it by Y' - one multiplies it by some more reasonable function then one gets still an "infinitesimal" result ; in this latter case one could as well have considered $Y^n = Y$, as classically. This examplifies the general fact that the theory of generalized functions in $\mathcal{G}(\Omega)$ can be considered as a _refinement of classical analysis_ , without any contradiction with classical analysis and distribution theory (as long as one considers only calculations valid inside distribution theory) ; the above example of multiplication by u in (1) does not make sense inside distribution theory. These calculations would a priori make sense in $\mathcal{G}$ but $Y^n \neq Y$ in $\mathcal{G}$ ($\mathbb{R}$) as soon as n $\neq$ 1. In view of that it appears that the assumption that the classical algebra $\mathcal{C}_f$ ($\mathbb{R}$) (of all piecewise constant functions) is a subalgebra of A (underlying in Schwartz's impossibility result, see § 1. 3) can be considered as irrealistic.

Research problem. The reader is assumed to know the definition of $\mathscr{G}(\Omega)$ and Nonstandard Analysis. Clarify the connections between our concept of generalized functions and the nonstandard functions. Since there is no canonical inclusion of $\mathfrak{D}'(\Omega)$ into the set of nonstandard functions Nonstandard Analysis is probably much closer to the simplified concept $\mathscr{G}_s(\Omega)$ defined below in § 8.4. Various constructions of Nonstandard Analysis mimicking the construction of $\mathscr{G}(\Omega)$ are given in Oberguggenberger [9] and Todorov [1]. Since both theories realize a differential and integral calculus dealing with infinitesimal and infinitely large quantities it seems to me that a fusion (of both theories) is perhaps possible.For a comparison of the two theories in the context of nonlinear hyperbolic equations see Oberguggenberger [12].

§1. 2 AN INTRODUCTION TO DISTRIBUTIONS. This section is intended to the reader who does not know distribution theory ; it can be dropped by the other readers. If Ω is an open set in $\mathbb{R}^n$ we denote by $\mathfrak{D}(\Omega)$ the vector space of all (scalar valued) $\mathscr{C}^\infty$ functions on Ω with compact support (such functions exist ! ; the support of a function f (denoted by supp f) is the closure of $\{x \mid f(x) \neq 0 \}$). We say that a sequence (f_n) of functions in $\mathfrak{D}(\Omega)$ "tends to 0" (notation "$f_n \to 0$") if and only if 1) and 2) below are satisfied :

   1) their supports are contained in a fixed compact subset of Ω
   2) for every partial derivative D (including the identity)

$$\lim_{n \to \infty} \sup_{x \in \Omega} |D f_n(x)| = 0.$$

Definition. A distribution on Ω is a linear form $T : \mathfrak{D}(\Omega) \to \mathbb{C}$ such that $T(f_n) \to 0$ in $\mathbb{C}$ as soon as "$f_n \to 0$".

We denote by $\mathfrak{D}'(\Omega)$ the space of all distributions on Ω. $\mathfrak{D}'(\Omega)$ is a vector space.

We define partial derivatives of distributions by : if $T \in \mathfrak{D}'(\Omega)$, $\dfrac{\partial T}{\partial x_i} \in \mathfrak{D}'(\Omega)$ is the distribution defined by $\dfrac{\partial T}{\partial x_i}(\varphi) = - T\left(\dfrac{\partial \varphi}{\partial x_i}\right) \quad \forall \varphi \in \mathfrak{D}(\Omega)$;

thus

$$DT(\varphi) = (-1)^{o(D)} T(D\varphi) \quad \forall \varphi \in \mathfrak{D}(\Omega)$$

if D is an arbitrary partial derivative operator and if o(D) is its order.

We multiply a $\mathscr{C}^\infty$ function and a distribution according to the formula : if $\alpha \in \mathscr{C}^\infty(\Omega)$ and $T \in \mathfrak{D}'(\Omega)$, the product $\alpha. T \in \mathfrak{D}'(\Omega)$ is defined by

$$(\alpha. T)(\varphi) = T(\alpha \varphi) \quad \forall \varphi \in \mathfrak{D}(\Omega).$$

Any locally integrable function is a distribution : if $f \in L^1_{loc}(\Omega)$ then we set

$$f(\varphi) = \int_{\Omega} f(x)\, \varphi(x)\, dx \qquad \forall\, \varphi \in \mathcal{D}(\Omega).$$

Since it is known that $f(\varphi) = 0 \;\; \forall\, \varphi \in \mathcal{D}(\Omega) \Rightarrow f = 0$ in $L^1_{loc}(\Omega)$ one has an inclusion $L^1_{loc}(\Omega) \subset \mathcal{D}'(\Omega)$. If $p = 2, 3, \ldots, \infty$ one has similarly an inclusion $L^p_{loc}(\Omega) \subset \mathcal{D}'(\Omega)$.

One checks at once that the differentiation of a distribution and the multiplication by a $\mathscr{C}^{\infty}$ function extend these respective classical operations (in $\mathscr{C}^1(\Omega)$ and in $L^p_{loc}(\Omega)$ respectively). However note that if f is a classical function which is twice differentiable (but not twice continuously differentiable) and such that $f''_{x,y} \neq f''_{y,x}$ (such functions exist !) then since $f''_{x,y} = f''_{y,x}$ in the sense of distributions, the classical and distributional second derivatives are not identical.

Example 1 : the Dirac delta distribution defined by the formula $\delta(\varphi) = \varphi(0)$. If $\delta^{\varepsilon} \in \mathscr{C}(\mathbb{R})$,

$0 < \varepsilon < 1$, $\delta^{\varepsilon} \geq 0$, $\int \delta^{\varepsilon}(x)\, dx = 1$, supp $\delta_{\varepsilon} \subset [-\varepsilon, +\varepsilon]$, then if $\varphi \in \mathcal{D}(\mathbb{R})$

$$\int_{\mathbb{R}} \delta^{\varepsilon}(x)\, \varphi(x)\, dx = \int_{\mathbb{R}} \delta^{\varepsilon}(x)\, (\varphi(0) + x\varphi'(\theta x))\, dx \to \varphi(0) \text{ as } \varepsilon \to 0 \; (0 < \theta < 1).$$

One says that $\delta^{\varepsilon} \to \delta$ in $\mathcal{D}'(\mathbb{R})$ as $\varepsilon \to 0$.

Example 2. the derivative of the Heaviside function : prove that the derivative Y' of the Heaviside function Y is the Dirac distribution δ.

One can prove (Schwartz [1])the structure theorem :
Theorem . Any distribution is locally a partial derivative of a continuous function.
In other words : $\forall\, T \in \mathcal{D}'(\Omega) \; \forall x_0 \in \Omega \; \exists$ an open neighborhood V_{x_0} of x_0 in Ω, $\exists f \in \mathscr{C}$ (V_{x_0}) and $\exists$ a partial derivation operator D such that

$$T|_{V_{x_0}} = Df \quad \text{in } \mathcal{D}'(V_{x_0})$$

where $T|_{V_{x_0}}$ is the restriction of T to V_{x_0} (obvious definition : one considers only the test functions $\varphi \in \mathcal{D}(V_{x_0}) \subset \mathcal{D}(\Omega)$).

From this structure theorem the distributions constitute the smallest space in which it is permitted to differentiate (infinitely) all continuous functions (and also all L^p_{loc} functions p = 1, 2, ..., ∞.).

Finally the distributions enjoy essentially all the nice properties of the $\mathcal{C}^\infty$ functions, with the basic exceptions of the <u>multiplication</u> (as well as all main nonlinear operations ; try to multiply "reasonably" Y and δ, δ and δ,..), of the <u>restriction to a vector space</u> (let δ_2 be the Dirac distribution on $\mathbb{R}^2$: δ_2 (φ) = φ (0,0) ; try to restrict δ_2 to $\mathbb{R} \times \{0\}$), and of <u>the composition product</u> (try to define the composition $f_o\delta$ (f ∈ $\mathcal{C}^\infty$ ($\mathbb{R}$)).

Various extensions of the distributions have been proposed.

<u>The ultradistributions</u>. They are defined by replacing $\mathcal{D}(\Omega)$ by a smaller space of $\mathcal{C}^\infty$ functions, satisfying for instance, in one dimension, bounds of the kind

$$\| \varphi^{(k)} \|_\infty \leq M\, C^k (k!)^s, s > 1$$

(for s = 1 the function φ is analytic and so cannot have compact support unless it is the zero function). Various spaces of ultra-distributions are defined as the duals of such subspaces of $\mathcal{D}(\Omega)$; these spaces contain $\mathcal{D}'(\Omega)$ but do not have very different properties ; see Lions-Magenes [1] for definitions and references.

<u>The analytic functionals.</u> One considers a space of analytic functions, for instance the space $\mathcal{H}(\Omega)$ of all holomorphic functions on an open set $\Omega \subset \mathbb{C}^n$, equipped with the topology of uniform convergence on the compact subsets of Ω. The space of analytic functionals is defined as the dual $\mathcal{H}'(\Omega)$. Since any analytic function with compact support is the constant 0 there are difficulties to define the support of an analytic functional ; further one can only multiply the analytic functionals by analytic functions. See Martineau [1].

<u>The hyperfunctions</u> generalize both the distributions and the analytic functionals, see Martineau [2]. Grosso modo a hyperfunction on $\mathbb{R}^n$ appears as a locally finite series of analytic functionals that patch together.

There are linear PDEs without distribution solutions, but that have solutions which are analytic functionals or hyperfunctions. However all these extensions of the distributions share essentially the same properties : unlimited differentiation but impossibility of the multiplication in general ; also, like in the setting of distributions, many very simple linear PDEs with polynomial coefficients do not have solutions in these spaces, see § 1. 4 below.

Research Problem. The reader is assumed to know the definition of $\mathcal{G}(\Omega)$; it is clear that this definition can be modified so as to include the ultra-distributions (and so to permit a general multiplication of ultra-distributions). Is it possible - probably at the price of a greater modification in the definition of $\mathcal{G}(\Omega)$ - to include the analytic functionals and / or the hyperfunctions in a differential algebra looking like $\mathcal{G}(\Omega)$ (thus permitting a general multiplication of analytic functionals and / or hyperfunctions)? A special type of ultradistributions has been included in a larger algebra in Gramchev [1].

§1. 3 SCHWARTZ IMPOSSIBILITY RESULT.

Theorem [Schwartz [2], 1954]. Let A be an algebra containing the algebra $\mathscr{C}(\mathbb{R})$ of all continuous functions on $\mathbb{R}$ as a subalgebra. Let us assume that the constant function $1 \in \mathscr{C}(\mathbb{R})$ is the unit element in A. Further let us assume that there exists a linear map $D : A \to A$ generalizing the derivation of continuously differentiable functions and satisfying Leibniz's rule ($D (a. b) = Da. b + a. Db$). Then one has $D^2 (|x|) = 0.$

Of course D (|x|) has values -1 for $x < 0$ and +1 for $x > 0$, therefore D^2 (|x|) should be null outside 0, "infinite" in 0 so that

$$\int_{-\infty}^{+\infty} D^2(|x|) \, dx = [D(|x|)]_{-\infty}^{+\infty} = 2.$$ Thus the conclusion of the theorem contradicts any reasonable

intuition. In distribution theory $D^2(|x|) = 2\delta$ and so the above result shows that A cannot contain the Dirac delta function, thus making the algebra A uninteresting.

Basic Remark. The distributions are not involved in Schwartz's impossibility result : the algebra $\mathscr{C}(\mathbb{R})$, the differentiation of continuously differentiable functions, and the usual calculation rules are the only ingredients that produce the impossibility. And all these ingredients are perfectly natural ! However it has already been noticed in §1.1 that the multiplication of piecewise constant functions together with the usual rules of differentiation produces at once a contradiction.

Before the proof we give a lemma.

Lemma In A $xa = 0 \Rightarrow a = 0$ (where x is the classical function $x \to x$ and where a is an arbitrary element of A).

Proof of the theorem.

$$D(x|x|) = Dx . |x| + x. D(|x|) = |x| + x.D(|x|).$$

Therefore

$$D^2(x|x|) = 2D(|x|) + xD^2(|x|).$$

In $\mathscr{C}^1(\mathbb{R})$, hence in A :

$$D(x|x|) = 2\,|x|.$$

Therefore

$$D^2(x|x|) = 2D|x|.$$

It follows from the two above expressions for $D^2(x|x|)$ that x. $D^2(|x|) = 0$ thus from the lemma $D^2(|x|)$ = 0

□

Proof of the lemma. The functions $x(\log |x| - 1)$ and $x^2(\log |x| - 1)$ are in $\mathscr{C}(\mathbb{R})$ provided we give them the value 0 for x = 0. Using Leibniz's rule in A

$D\{x(\log |x| - 1).x\} = D\{x(\log|x| - 1)\}.x + x(\log|x| - 1)$
$D^2\{x(\log|x| - 1).x\} = D^2\{x(\log |x| - 1)\}.x + 2\,D\{x(\log|x| - 1)\}.$
Thus
(8) $D^2\{x(\log|x| - 1)\}.x = D^2\{x^2(\log|x| - 1)\} - 2D\{x(\log|x| - 1)\}.$

But, since D coincides with the usual derivation operator on $\mathscr{C}^1$ functions and since the function x^2 $(\log |x| - 1)$ is a $\mathscr{C}^1$ function :

$D\{x^2(\log|x| - 1)\} = 2x(\log|x| - 1) + x.$

Therefore in A
(9) $D^2\{x^2(\log|x| - 1)\} = 2D\{x(\log|x| - 1)\} + 1.$

(8) and (9) yield :

$$D^2\{x(\log|x| - 1)\}.x = 1.$$

To simplify the notation set $y = D^2\{x(\log|x| - 1\}$; then y.x = 1; thus x.a = 0 $\Rightarrow$ y (xa) = 0 $\Rightarrow$ (yx)a = 0 $\Rightarrow$ 1.a = 0 $\Rightarrow$ a= 0.

□

A more detailed discussion is given in Rosinger [1] Part I chap 2.

Research problem. Many particular multiplications of distributions have been considered, see Colombeau [1] chap 2, Rosinger [1] App. 5 in Part 2, Oberguggenberger [11]. Up to now it has been proved that nearly all of them are particular cases (modulo some concept of "infinitesimality" as for the product of continuous functions, see chapter 8) of the multiplication in $\mathcal{G}(\Omega)$, see Oberguggenberger [1], Jelinek [1,2]. There remains some possible studies in this field.

§1.4 LINEAR PDEs WITHOUT DISTRIBUTION SOLUTION.

It is immediate to show that certain Cauchy problems do not have solution : for instance the equation

$$\begin{cases} (\frac{\partial}{\partial t} + i\frac{\partial}{\partial x})u = 0 \\ u(x,0) = u_0(x) \end{cases}$$

cannot have C^1(and also any distribution) solution in an open set Ω intersecting the line $t = 0$ if u_0 is $\mathcal{C}^\infty$ but not analytic. Indeed a solution u would be holomorphic in Ω(one can prove that if $u \in \mathcal{D}'(\Omega)$, $\Omega \subset \mathbb{C}$ open, then $\frac{\partial}{\partial \bar{z}} u = 0 \Rightarrow u \in \mathcal{H}(\Omega)$). Therefore u_0 would be a real analytic function.

The above equation cannot also have solutions in $\Omega \cap t > 0$, the initial condition being understood as a limit when $t \to 0$: extend this solution to $t < 0$ by setting $u(x,t) = \overline{u(x,-t)}$ (where the bar denotes complex conjugation) and apply the proof above. But one can prove (Hörmander [1]) that for any $\mathcal{C}^\infty$ function f on $\mathbb{R}^2$ there is $u \in \mathcal{C}^\infty(\mathbb{R}^2)$ such that $(\frac{\partial}{\partial t} + i\frac{\partial}{\partial x})$ u = f. Considerable effort has been invested on the following problem. Let

$$P(x,D) = \sum_{\substack{p \in \mathbb{N}^n \\ |p| \le m}} c_p(x) D^p, \quad (c_p \in \mathcal{C}^\infty(\mathbb{R}^n), D^p = \frac{\partial^{|p|}}{\partial x_1^{p_1}...\partial x_n^{p_n}})$$

be a nonzero linear partial differential operator with $\mathcal{C}^\infty$ coefficients.

Problem. Let $f \in \mathcal{C}^\infty(\mathbb{R}^n)$ and let $x_0 \in \mathbb{R}^n$ be given ; is there V_{x_0}, open neighborhood of x_0, and $u \in \mathcal{D}'(V_{x_0})$ such that

$$P(x,D) u = f \quad \text{in } V_{x_0}?$$

H. Lewy [1] has produced a celebrated counterexample. Let $x_1, x_2, y_1 \in \mathbb{R}$ and let us consider the equation

(L)
$$\left[-\frac{\partial}{\partial x_1} - i\frac{\partial}{\partial x_2} + 2i(x_1 + ix_2)\frac{\partial}{\partial y_1}\right]u = f(y_1).$$

Then for any f which is $\mathscr{C}^\infty$ and not analytic, and any point $x_o \in \mathbb{R}^3$ equation (L) provides a negative answer to the above problem.

<u>Sketch of proof :</u> for simplification we shall only consider the case $x_o = (0,0,y_1^0)$ and the case u is a $\mathscr{C}^1$ function (easy extension to the case u is a distribution). Set $x_1 + ix_2 = y_2^{1/2} e^{i\theta}$, $y_2 > 0$ and let

$$U(y_1,y_2) = i\int_0^{2\pi} e^{i\theta} y_2^{1/2} u(x_1,x_2,y_1) d\theta.$$

From (L) one computes (Lewy [1]) that

$$\frac{\partial U}{\partial y_1} + i\frac{\partial U}{\partial y_2} = \pi f(y_1).$$

Let F be real such that $F' = f$; then the function

$$V(y_1,y_2) = U(y_1,y_2) - \pi F(y_1)$$

is $\mathscr{C}^1$ and satisfies

$$\frac{\partial V}{\partial y_1} + i\frac{\partial V}{\partial y_2} = 0$$

in the intersection of an open ball in the (y_1,y_2) plane centered at $(y_1^0, 0)$ and the half plane $y_2 > 0$; thus it is holomorphic in this upper half ball. Further $U(y_1, 0) = 0$ and $V(y_1,0) = -\pi F(y_1)$ and so V is real valued on $y_2 = 0$. Thus V can be continued holomorphically in the whole of the open ball ; since $F(y_1) = -\frac{1}{\pi} V(y_1,0)$ F is an analytic function and so f is analytic. $\square$

A similar counterexample has been given in the space of hyperfunctions (larger than the space of distributions), see Shapira [1]. Since then a great amount of work has been devoted to the research of necessary and of sufficient conditions for local solvability, see Hörmander [2].

More details on the contents of this section, and other equations without solution, are given in Rosinger [1] Part1 chap.3.

Research problem. It has been proved that also in $\mathcal{G}$ linear PDEs may fail to have solutions. Thus the problem is to find - still in $\mathcal{G}$ or in a similar setting of generalized functions - a convenient formulation of the PDEs, even the linear ones with $\mathcal{C}^\infty$ coefficients, allowing general existence results. Of course coherence with the classical solutions - when they exist - should be obtained. An attempt is presented in Rosinger [1] Part II chap 3, Egorov [1], Colombeau-Heibig-Oberguggenberger [1].

Chapter 2. Multiplications of distributions in classical physics.

§2. 1 <u>ELASTICITY AND ELASTOPLASTICITY.</u> In this section we consider large deformations of solid bodies, that could be produced for instance by a strong collision. These large deformations may lead to plastic or other forms of structural failure. At the level of numerical computations this imposes a Eulerian description (i. e. with a fixed frame of reference) since the Lagrangian description (i. e. with a frame of reference following the deformations of the medium) is subject to numerical failure at large deformations. Experience has shown that Eulerian methods can work very well. The system of equations modeling the behaviour of solids includes at first the basic classical laws of conservation of mass, momentum and energy ; usually viscosity is neglected. The basic conservation laws are completed by "constitutive equations" obtained from experiments on each material. The constitutive equations can take very different forms (they distinguish steal from rubber since the conservation laws are the same !). In this text we limit ourselves to the simplest models of elasticity and elastoplasticity, which can be stated as follows (see Arnaud [1] for instance). At first we begin with the purely elastic case.

<u>Notation.</u> $x = (x_1, x_2, x_3) =$ space coordinate, $t =$ time .

$\rho =$ density

$\vec{U} = (u_1, u_2, u_3) =$ velocity vector

$\overline{\Sigma} =$ stress tensor, with components $\sigma_{i,j}, 1 \le i,j \le 3$

$p = -\frac{1}{3}$ trace $(\overline{\Sigma}) =$ pressure

$\mathbb{1} =$ identity 3 x 3 matrix

$\overline{S} = \overline{\Sigma} + p \mathbb{1} =$ stress deviation tensor

$\overline{V} =$ rate of deformation tensor, of components $v_{i,j} = \frac{1}{2} \left(\frac{\partial u_i}{\partial x_j} + \frac{\partial u_j}{\partial x_i} \right) 1 \le i, j \le 3$

$\overline{\Omega} =$ spin tensor, of components $\omega_{i,j} = -\frac{1}{2} \left(\frac{\partial u_i}{\partial x_i} - \frac{\partial u_j}{\partial x_j} \right)$

$e =$ total specific energy

$I = e - \frac{1}{2} \vec{U} . \vec{U} =$ internal specific energy

$\frac{d}{dt}$ denotes the particle time derivative : if f is a function of (x,t) then

$$\frac{d}{dt} f = \frac{\partial f}{\partial t} + \vec{U} . \overrightarrow{grad} f$$

if $\overline{A} = (a_{ij})$ is a 3 x 3 tensor we set

$$(\mathrm{div}\ \overline{A}\ )_i = \sum_j \frac{\partial a_{ij}}{\partial x_j}\ ;$$

then div $\overrightarrow{A}$ is a vector.

If $\overrightarrow{V} = (v_1, v_2, v_3)$ is a vector the tensor $\overline{A} = \mathrm{Grad}\ \overrightarrow{V}$ is defined by $a_{ij} = \frac{\partial v_i}{\partial x_j}$. If $\overline{A}$ and $\overline{B}$ are

tensors the tensor $\overline{C} = \overline{A}\ .\ \overline{B}$ is defined by $c_{ij} = \sum_k a_{ik}b_{kj}$. If $\overline{A}$ is a tensor and $\overrightarrow{V}$ is a vector the

product $\overrightarrow{U} = \overline{A}\ \overrightarrow{V}$ is the vector defined by $u_i = \sum_k a_{ik}v_k$. If $\overline{A}$ is a tensor sym $\overline{A}$ denotes its

symmetric part.

      With this set of notation a Eulerian model of pure elasticity used by engineers is the system of equations :

(1) $\begin{cases} \dfrac{d}{dt}\rho + \rho\ \mathrm{div}\ \overrightarrow{U} = 0 \ \ \text{mass conservation} \\[2mm] \rho\ \dfrac{d}{dt}\ \overrightarrow{U} = \overrightarrow{\mathrm{div}}\ \overline{\Sigma}\ \text{momentum conservation} \\[2mm] \rho\ \dfrac{d}{dt}e = \mathrm{div}\ (\ \overline{\Sigma}\ \overrightarrow{U}\ )\ \ \text{energy conservation} \\[2mm] p = \phi(\rho,I)\ \ \text{equation of state} \\[2mm] \dfrac{d}{dt}\ \overline{\Sigma} = 2G\overline{V} + \lambda(\mathrm{trace}\ \overline{V})\ \mathbb{1} + \mathrm{sym}\ (\overline{\Omega}.\overline{\Sigma}\ )\ \ \text{(G is called the shear modulus) Hooke's law.} \end{cases}$

The last equation follows from Hooke's law (linear stress - strain relationship) in a Lagrangian frame, and then from the Lagrange - Euler change of coordinates ; G is called the shear modulus ; λ is a Lamé constant. From the relation $\overline{\Sigma} = \overline{S} - p\,\mathbb{1}$ this last equation can be divided into the isotropic and deviation parts :

(2) $\begin{array}{l} a \\ \\ b \end{array} \begin{cases} \dfrac{d}{dt}p = - (3\lambda + 2G)\ \mathrm{div}\ \overrightarrow{U} \\[3mm] \dfrac{d}{dt}\overline{S} = 2G\ [\overline{V} - \tfrac{1}{3}(\mathrm{trace}\ \overline{V}\ )\ \mathbb{1}\ ] + \mathrm{sym}\ (\overline{\Omega}\ .\ \overline{S}\ ). \end{cases}$

Usually equation 2 a) is dropped (considered as redundant with the equation of state in (1)). These equations are written in case of smooth flows, but their solution shows shock waves and they are also adopted in case of shock waves. It has been observed that certain discretizations (the Hull code Duret - Matuska [1] , Matuska [1], for instance) give good numerical results, in agreement with those from observations.

For more simplicity we shall restrict our discussion to the 2 dimensional case. Then system (1) (in which 2a) has been dropped) reduces to the following system (8 unknown functions, 8 equations) in which the following notation has been adopted :
space coordinates (x,y); $G = \mu$: another Lamé constant ; velocity components in the x,y directions respectively : (u,v) ; $\overline{S} = (s_{i,j})$, $1 \leq i, j \leq 2$; $\quad \sum = (\sigma_{i,j})_{1 \leq 1, j \leq 2}$

(3)
$$
\begin{cases}
\rho_t + (\rho u)_x + (\rho v)_y = 0 \quad \text{mass conservation} \\[4pt]
\left.\begin{aligned}
&(\rho u)_t + (\rho u^2)_x + (\rho u v)_y + (p - s_{11})_x - (s_{12})_y = 0 \\
&(\rho v)_t + (\rho u v)_x + (\rho v^2)_y + (p - s_{22})_y - (s_{12})_x = 0
\end{aligned}\right\} \begin{aligned}&\text{momentum}\\&\text{conservation}\end{aligned} \\[10pt]
(\rho e)_t + (\rho e u)_x + (\rho e v)_y + [(p - s_{11})u]_x + [(p - s_{22})v]_y - (s_{12}v)_x - (s_{12}u)_y = 0 \\
\text{energy conservation} \\[10pt]
\left.\begin{aligned}
&(s_{11})_t + u(s_{11})_x + v(s_{11})_y = \tfrac{4}{3}\mu u_x - \tfrac{2}{3}\mu v_y + (u_y - v_x)s_{12} \\
&(s_{22})_t + u(s_{22})_x + v(s_{22})_y = -\tfrac{2}{3}\mu u_x + \tfrac{4}{3}\mu v_y - (u_y - v_x)s_{12} \\
&(s_{12})t + u(s_{12})_x + v(s_{12})_y = \mu v_x + \mu u_y - (u_y - v_x)\tfrac{s_{11} - s_{22}}{2}
\end{aligned}\right\} \text{Hooke's law} \\[10pt]
p = \phi\,(\rho, I) \quad \text{equation of state}
\end{cases}
$$

In 1 dimension this system gives ($v = 0 = s_{12} = u_y = p_y$, s_{22} given by an independent equation) the system of 5 equations for the 5 unknown functions ρ, u, p, s, e :

(4)
$$
\begin{cases}
\rho_t + (\rho u)_x = 0 \quad \text{mass conservation} \\
(\rho u)_t + (\rho u^2)_x + (p - s)_x = 0 \quad \text{momentum conservation} \\
(\rho e)_t + [\rho e u + (p - s)u]_x = 0 \quad \text{energy conservation} \\
s_t + u s_x - \tfrac{4}{3}\mu u_x = 0 \quad \text{Hooke's law} \\
p = \phi\,(\rho, I) \quad \text{equation of state}
\end{cases}
$$

Systems (3) and (4) are those adopted in Cauret [1].

Definition : a system of equations is said to be in conservation from (or a conservative system) if it is stated in the form

(5) $\quad \vec{U}_t = \mathrm{Div}\, f(\vec{U})$

for a certain function f, where $\vec{U}$ is the column of the unknown functions.

In (3) the equations expressing Hooke's law are in nonconservation form because of the terms of the kind us_x and u_xs (the single term us_x in (4)). In case of a shock wave the unknown functions are discontinuous on the shock : in particular u and s are simultaneously discontinuous and so the term $u\,s_x$ appears in the form of the meaningless product $Y.\,\delta$ of the Heaviside function and the Dirac function.

These equations model elasticity. Elastoplasticity is as follows : when the modulus of the constraint is large enough (in one dimension we shall consider for simplification that this modulus is $|s|$; in several dimensions this might be $\|S\| = (\mathrm{trace}\,(S^2))^{1/2}$, see Arnaud [1]) then Hooke's law ceases to be valid and the material becomes plastic, i.e. it behaves like a fluid, with $\|S\|$ remaining constant at its critical value. This phenomenon is called elastoplasticity. Then the Lamé constants μ and λ depend on $\|S\|$. In a simplified one dimensional model one can set $\mu(s) = \mu$ (Lamé constant) if $|s| \leq s_0$ (this is the purely elastic case) and $\mu(s) = 0$ if $|s| = s_0$ (then $|s|$ cannot reach values strictly larger than s_0 ; this is the plastic case). Note that the material can be purely elastic at the beginning of a shock, then reach the plastic stage in which very large deformations usually occur and then go back to the purely elastic case (in a shape quite different from the one it had at the beginning). Indeed this is the usual picture of a (strong enough) collision of two solid materials. The function μ $(|s|)$ can be discontinuous (as described above) or continuous : in all cases the term $\mu(|s|)\,u_x$ in (4) (and the numerous similar terms in 2 and 3 dimension) gives rise to a meaningless product of the kind $Y.\,\delta$.

Numerical methods have been developed, on an intuitive basis, for the solution of the above systems. In spite of the mathematical impossibility of discontinuous solutions (within distribution theory) they show such discontinuous solutions (shock waves) and give very good results ; they are widly used in industry (for the numerical simulation of collisions, to design armour for instance). Numerical results are given in Cauret [1] for 1 and 2 D. problems, in Arnaud [1] for 3 D. problems. Three of them are reproduced below (1D, 2D and 3D) to show the evidence of shock wave solutions (fig. 1) and the natural character of the results so obtained (they agree with results of experiments).

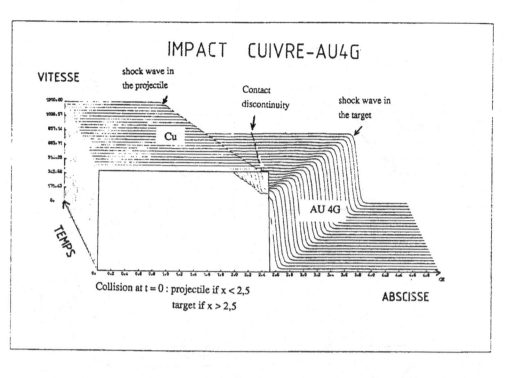

Figure 1. A numerical solution of system (4) (Cauret [1]) representing a collision of a projectile of copper (on the left) on a target of AU 4G (on the right). One observes a shock wave in the projectile propagating to the left, a shock wave in the target propagating to the right, and a contact discontinuity (the values of the densities being different on both sides of it). The figure depicts the values of the velocity u(x,t) in function of the space and the time. AU4G is an alloy from aluminium.

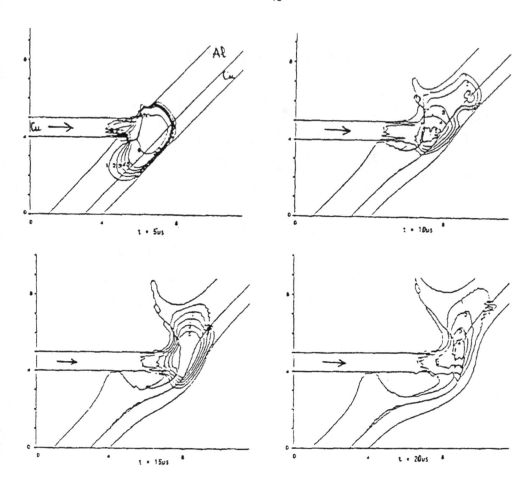

Figure 2 : A numerical solution of system (3) in two space dimensions (Cauret [1]). A half plane of copper coming from the left with a velocity of 3000 meters per second collides with a target made of one layer of aluminium and one layer of copper. The lines are isopressure lines; Collision begins at t = 0, figures are given at t = 5, 10, 15, 20 microseconds. This shock produces shock waves, which however are not apparent in the present description with isopressure lines.

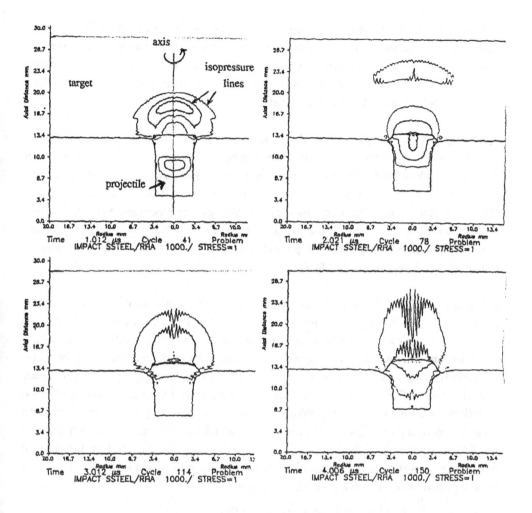

<u>Figure 3</u> : A 3D numerical solution of system (1) in the axisymmetric case from the Hull code. One observes severe pressure oscillations due to a bad numerical treatment, partly repaired in Arnaud [1].

Remark 1 : In the case of pure elasticity a conservative model has been proposed in Plohr - Sharp [1]. Indeed the nonconservative character of the above systems is due to the statement of Hooke's law. Since various constitutive relations may be proposed it is very hard to know whether the conservative or the nonconservative ones will be retained by physicists ; probably both will be used. As far the A. knows conservative models have not been found in elastoplasticity.

Problem 2. A detailed study (in the purely elastic case as well as in the elastoplastic case) has been carried on within the framework of generalized functions proposed in this book, see the sequel. Various research problems will be exposed on this theme. In particular, since it is in conservative form the Plohr - Sharp system of elasticity has nonambiguous jump conditions. Soon we shall show that systems in nonconservation form have usually ambiguous jump conditions. Which of these ones (in (1),(3),(4) or similar systems in use) is equal (or closer to) the Plohr Sharp jump conditions ? Then compare numerical methods (in particular compare nonconservative methods) with the conservative methods on the Plohr - Sharp system.

§ 2. 2 MULTIFLUID FLOWS.

In this section we consider a mixture of two fluids that are interpenetrating : each one has its own velocity vector field, and all their respective velocity fields exist at all points. We assume the fluids do not mix on a molecular level, as would gases. We assume that at any time the spatial regions occupied by the various fields are disjoint. Interpenetration arises as an approximation on a scale coarser than the molecular scale. For instance our flow could be made of bubbles of one fluid scattered into the other. Since the geometric configuration could be too complicated we have to treat the problem by some kind of averaging procedure in which the geometric configuration is ignored. We also assume the fluids do not interact with each other (by some chemical procedure for instance, ...) and so they are only governed by the classical conservation laws and constitutive equations. Let us consider n fluids and let α_i, $1 \leq i \leq n$, be the volumic proportion of the ith fluid in the mixture . Thus

$$(6) \qquad \sum_{i=1}^{n} \alpha_i (x,y,z,t) = 1 \text{ and } \alpha_i (x,y,z,t) \geq 0.$$

A great variety of models have been proposed, see Stewart - Wendroff [1]. One that works well is the following (the "standard model" in Stewart - Wendroff [1]). Let ρ_i denote the density of the ith fluid in the mixture, $\vec{V}_i$ its velocity vector, e_i and I_i its total and internal energy densities respectively: $e_i = I_i + \frac{1}{2} \vec{V}_i . \vec{V}_i$. We denote by p the pressure of the mixture (as a resultant of the pressures produced by each fluid). The model is

$$(7)\begin{cases} (\alpha_i\rho_i)_t + \text{Div } (\alpha_i\rho_i \ \vec{V}_i) = 0 \\ (\alpha_i\rho_i \ \vec{V}_i)_t + \overrightarrow{\text{Grad}}(\alpha_i\rho_i \ \vec{V}_i.\vec{V}_i) + \alpha_i. \ \overrightarrow{\text{Grad}} \ p = 0 \\ (\alpha_i \ \rho_i \ e_i)_t + \text{Div } (\alpha_i.(\rho_i \ e_i + p). \ \vec{V}_i) + p.(\alpha_i)_t = 0 \\ p = \phi_i(\rho_i, I_i). \end{cases}$$

Of course the three first equations are formulations of the conservation laws of mass, momentum and energy, and the last equation is made of the state laws of the various fluids. (6) (7) give 6n+1 equations for 6n+1 unknowns. Shock waves can be observed in the mixture (we assume each fluid is inviscid). Then the terms $\alpha_i \ . \ \overrightarrow{\text{Grad}} \ p$ and $p.(\alpha_i)_t$ appear in the form of meaningless products of distributions of the kind Y. δ. In contrast with the case of §2.1 the products of distributions do not arise from the constitutive equations but from some kind of simplification to avoid the consideration of the (possibly very complicated) geometry of the system. In the case of two fluids and in the simplified case in which the state laws involve only the pressure (and not the internal energy) one gets in 1 D the following system

$$(8)\begin{cases} (\alpha\rho_1)_t + (\alpha\rho_1 u_1)_x = 0 \\ ((1-\alpha)\rho_2)_t + ((1-\alpha)\rho_2 u_2)_x = 0 \\ (\alpha\rho_1 u_1)_t + (\alpha\rho_1 u_1^2)_x + \alpha p_x = 0 \\ ((1-\alpha)\rho_2 u_2)_t + ((1-\alpha)\rho_2 u_2^2)_x + (1-\alpha)p_x = 0 \\ \rho_1 = \rho_1(p) \ , \ \rho_2 = \rho_2(p) \end{cases}$$

in which u_1 and u_2 are the respective velocities of the fluids. Multiplications of distributions are in the term αp_x. These equations will be studied in the sequel of this book; here are numerical tests that show the evidence of shock waves for the systems above.

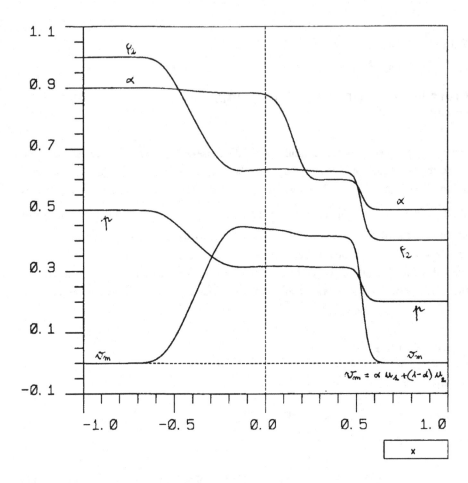

Figure 4 : a numerical solution of system (8) from a very simple scheme described in 3. 4. 1. below ;
on the right hand side one observes a shock wave in which all variables vary simultaneously.

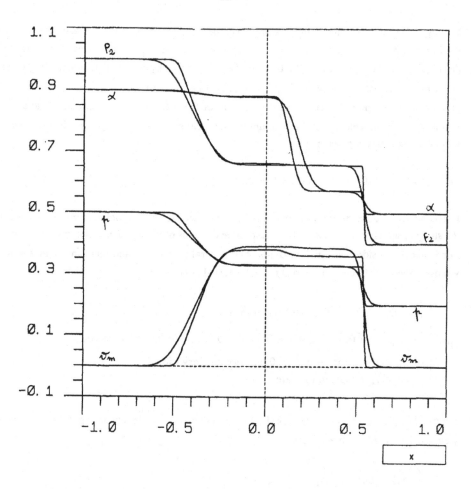

Figure 5 : a comparison of two numerical solutions of a system very close to (8), given in Liles-Reed [1] : the curves with steep shocks are obtained from the numerical scheme in Liles-Reed [1](Trac code), the diffusive curves are obtained from the very simple (but diffusive) scheme of the kind of the one in 3. 4. 1 below.

§2.3 ACOUSTICS. In this section we are concerned with the propagation of acoustic waves in a medium with piecewise $\mathscr{C}^\infty$ characteristics. The equations of acoustics are obtained from a perturbation of the equations of fluids or solids. Let w be a physical variable (depending on space and time), for instance $w = \rho, \vec{u}, p,....$ In acoustics one introduces a "small" dimensionless parameter ε (that may be defined as the quotient of the maximum of amplitude of the acoustics perturbation of the velocity of the medium by the local sound speed). One develops the physical variables in a series of powers of ε :

$$(9) \qquad w(x,t) = w_0(x,t) + \varepsilon\, w_1(x,t) + \varepsilon^2 w_2(x,t) + ... \, .$$

Linear acoustics consists in the study of the first order terms, weakly nonlinear acoustics consists in the study of terms of order 2 in ε. Their respective domains of validity are well known by physicists. For instance consider acoustics in fluids. The general system of fluid mechanics (in absence of viscosity, thermal effects and external forces) is (usual notations)

$$(10) \begin{cases} \rho_t + \text{Div}\,(\rho\,\vec{u}) = 0 \quad \text{mass conservation} \\ (\rho\,\vec{u})_t + \overrightarrow{\text{Grad}}\,(\rho\,\vec{u}\cdot\vec{u}) + \overrightarrow{\text{Grad}}\,p = 0 \quad \text{momentum conservation} \\ (\rho e)_t + \text{Div}\,(\rho e\,\vec{u} + p\vec{u}) = 0 \quad \text{energy conservation} \\ p = \phi(\rho,I) \quad \text{equation of state} \end{cases} .$$

Let $\rho_1, \vec{u}_1, p_1, ...$ denote the respective (first order) acoustic variables according to (9) . (9) and (10) yield the system of linear acoustics

$$(11) \begin{cases} (\rho_1)_t + \text{Div}\,(\rho_0\vec{u}_1 + \rho_1\vec{u}_0) = 0 \\ \cdot \\ \cdot \\ \cdot \end{cases} .$$

This is a linear system whose coefficients depend on $\rho_0,\ \vec{u}_0,\ p_0,...$ which are the physical variables of the medium free from acoustic perturbations. The system giving $w_2 = \rho_2,\ \vec{u}_2,\ p_2,\ ..$ (weakly nonlinear acoustics) is also a linear system ; its coefficients depend on w_0 and w_1. In the case of a medium with piecewise $\mathscr{C}^\infty$ characteristics (i. e. a juxtaposition of two immiscible fluids, of a fluid and a solid, ...) the coefficients $w_0 = \rho_0, \vec{u}_0, p_0,...$ are discontinuous. The acoustic variables w_i are also, in general, discontinuous on the discontinuities of the medium (even they often exhibit singularities of the kind of the Dirac delta function on these discontinuities). We shall ascertain that this gives rise to meaningless multiplications of distributions (of the kind Y. δ, or δ^2) on the surfaces of discontinuity of the medium (such a surface is called a dioptra).

One can work in Eulerian coordinates or in Lagrangian coordinates. In Eulerian coordinates the system of linear acoustics (Poirée [1,2, 3, 4]) is the following (case of a perfect fluid initialy at rest) Notation : s = entropy (s_0 = entropy of the medium without acoustic perturbation and s_1 = acoustic entropy), $\vec{u}_1$, p_1, s_1 are the respective <u>Eulerian</u> variables.

$\vec{\xi}_1$ is the acoustic displacement : if $\vec{a}$ is a particle (lagrangian coordinates)

$\vec{x}(\vec{a}, t) = \vec{a} + \varepsilon \vec{\xi}_1 (\vec{a}, t)$ where $\vec{x}$ is the position at time t of the particule $\vec{a}$

(12) $\begin{cases} (\rho_1)_t + \text{Div} (\rho \vec{u}_1) = 0 \text{ mass conservation} \\ (\rho_0 \vec{u}_1)_t + \overrightarrow{\text{Grad}} \, p_1 = 0 \text{ momentum conservation} \\ (s_1)_t + \vec{u}_1 \cdot \overrightarrow{\text{Grad}} \, s_0 = 0 \text{ constancy of entropy (in Lagrangian configuration)} \\ (\vec{\xi}_1)_t = \vec{u}_1 \text{ definition of the displacement} \end{cases}$

completed by the following complicated equation of state

(13) $\quad p_1 = c_0^2 \cdot \{\rho_1 + [\rho_0]_{\Sigma_0} \xi_{1, n_0} \, \delta(\lambda)\} + \alpha_0 \{s_1 + [s_0]_{\Sigma_0} \xi_{1, n_0} \, \delta(\lambda)\} -$

$\vec{\xi}_1 \cdot \{\overrightarrow{\text{Grad}} \, \mathcal{P} - [\mathcal{P}] \, \vec{n_0} \, \delta(\lambda)\}.$

The notation in (13) should be explained :

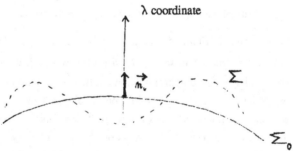

λ coordinate

Σ_0 is a dioptra without acoustic perturbation, Σ is the dioptra with acoustic perturbation, $\vec{n_0}$ is the unitary normal vector to Σ_0, $\xi_{1,n_0} = \vec{\xi}_1 \cdot \vec{n_0}$ (= normal component of the displacement vector), $\mathcal{P}$ comes from the equation of state of the fluid ; $p = \mathcal{P}(\rho, s_0; \vec{a}) = \mathcal{P}$, $c_0^2 = \dfrac{\partial \mathcal{P}}{\partial \rho}(\rho_0, s_0; \vec{a})$,

$\alpha_0 = \dfrac{\partial \mathcal{P}}{\partial s}(\rho_0, s_0; \vec{a})$, $[w_0]_{\Sigma_0}$ is the jump of w_0 on Σ_0 (w_0 is assumed not to be more irregular than discontinuous on Σ_0, and $\mathscr{C}^\infty$ outside Σ_0), λ is the coordinate on the normal line with unit vector

$\overrightarrow{n_0}$, δ is the Dirac delta function. (13) contains products of distributions of the kind Y. δ. Note that the formula (Poirée [2] p34)

(14) $w^L = w^E + \overrightarrow{\xi} . \overrightarrow{Grad} w_0$

(w^L = Lagrangian w^1, w^E = Eulerian w^1) shows that if w_0 and w^L are discontinuous functions on the dioptra (this is usually the case) then w^E contains a singularity in from of a Dirac function on the dioptra.

Since the equation of state is very complicated in the Eulerian description one prefers to use the Lagrangian description, Poirée [1,2,3, 4]. From now on $w^1 = \rho_1, \overrightarrow{u_1}, p_1,...$ will be Lagrangian acoustic variables, and $\overrightarrow{x}$ will be the Lagrangian space variable ; confusion with the notation in (12) (13) should be carefully avoided. One has the simple system (Poirée [1,2,3]) (notation ρ, $\overrightarrow{u}$,p for $\rho_1, \overrightarrow{u_1}, p_1$)

(15) $\begin{cases} \rho_t + \rho_0 \ Div \ \overrightarrow{u} = 0 \ \text{ mass conservation} \\ \rho_0 \overrightarrow{u}_t + \overrightarrow{Grad} \ p = 0 \text{ momentum conservation} \\ p = c_0^2 \ \rho \ \text{ equation of state} \end{cases}$

ρ_0 (x) and c_0^2(x) are C^∞ functions outside the dioptra, discontinuous on the dioptra (density and sound speed of the unperturbed medium). Multiplications of distributions of the kind Y. δ arise from the term $\rho_0 Div \ \overrightarrow{u}$ (at least at first sight, see 6.1.7 - 6.1.9 below ; such multiplications cannot be avoided in the second order acoustic approximation , called weakly nonlinear acoustics). Since Eulerian acoustic variables involve Dirac functions on the dioptra then the Eulerian equations of weakly nonlinear acoustics are linear equations whose coefficient functions contain Dirac functions. This gives examples of linear systems of equations of physics with very irregular coefficients : indeed putting (9) into (10) the mass conservation equation $\rho_t + (\rho u)_x = 0$ gives at the second order in ε the equation

(11') $(\rho_2)_t + (\rho_0 u_2 + \rho_1 u_1 + u_0 \rho_2)_x = 0,$

$\rho_0 = \rho_0(x, t)$ and $u_0 = u_0(x, t)$ are the given density and velocity of the medium in evolution free from acoustic perturbation (at rest in the simpler case), $\rho_1 = \rho_1$ (x, t) and $u_1 = u_1$ (x, t) are the first order acoustic density and pressure given as solutions of (11). Since ρ_0 and u_0 may be discontinuous

(11') is a linear equation involving multiplication of distributions and producing solutions in form of delta functions, see a study of the simplified equation

$$\frac{\partial}{\partial t} u(x, t) + \frac{\partial}{\partial x}(Y(x) u(x, t)) = 0$$

in 6.1.10, 6.1.11 below, Y(x) the Heaviside function.

Remark. Physicists are used to compute formally on these equations, Poirée [1, 2, 3, 4]. In contrast with the case of the equation $u_t + uu_x = 0$ (§1.1) or with the (much more technical) case of interacting field theory (§7.3;§7.4), no "catastrophe" appears in the calculations. This remark will be clarified later (6.1.12).

In the more general case in which the medium is made of a juxtaposition of elastic solids and fluids one uses the system of equations (in the Lagrangian description)

$$(16) \begin{cases} \vec{\xi}_t - \vec{u} = 0 & \text{definition of the displacement} \\ \rho_t + \rho_0 \, \text{Div} \, \vec{u} = 0 & \text{mass conservation} \\ \rho_0 \vec{u}_t - \overrightarrow{\text{Div}\overline{\Sigma}} = 0 & \text{momentum conservation} \\ \overline{\Sigma} = \lambda \, \text{Div} \, \vec{\xi} \, \mathbf{1} + \mu(\overrightarrow{\text{Grad} \, \vec{\xi}}) + (\overrightarrow{\text{Grad} \, \vec{\xi}})^T & \text{equation of state} \end{cases}$$

where $\overline{\Sigma}$ is the stress tensor, T the transposition symbol, λ and μ the Lamé constants ; in case some part of the medium is a fluid then one sets $\lambda = c_0^2 \rho_0$ and $\mu = 0$;the Lamé constants λ and μ depend on x and are discontinuous on the dioptra. Elimination of $\vec{\xi}$ gives the system

$$(16') \begin{cases} \rho_0 \partial_t \vec{u} - \overrightarrow{\text{Div}\overline{\Sigma}} = 0 \\ \partial_t \overline{\Sigma} - \lambda \, \text{Div} \, \vec{U} \, \mathbf{1} - \mu (\overrightarrow{\text{Grad} \, \vec{U}} + (\overrightarrow{\text{Grad} \, \vec{U}})^T) = 0. \end{cases}$$

In two space dimensions with notation $\vec{U} = (u_1, u_2)$, $\overline{\Sigma} = (\Sigma_{ij})_{1 \le i,j \le 2}$ with $\Sigma_{12} = \Sigma_{21}$ one obtains the system

$$(16'')\begin{cases} \rho_o \, \partial_t \, u_1 - \partial_x \, \Sigma_{11} - \partial_y \, \Sigma_{12} = 0 \\ \rho_o \, \partial_t \, u_2 - \partial_x \, \Sigma_{12} - \partial_y \, \Sigma_{22} = 0 \\ \partial_t \, \Sigma_{11} - (\lambda + 2\mu) \partial_x \, u_1 - \lambda \, \partial_y \, u_2 = 0 \\ \partial_t \, \Sigma_{22} - \lambda \, \partial_x \, u_1 - (\lambda + 2\mu) \, \partial_y \, u_2 = 0 \\ \partial_t \, \Sigma_{12} - \mu(\partial_x \, u_2 - \partial_y \, u_1) = 0. \end{cases}$$

The longitudinal and transverse wave speeds c_ℓ and c_t are related to the Lamé constants λ and μ by $c_\ell^2 = \dfrac{\lambda + 2\mu}{\rho_o}$ and $c_t^2 = \dfrac{\mu}{\rho_o}$. Fluids can be considered as elastic solids for which $\mu = 0$. Numerical solutions of $(16'')$ are given in 6.2.3 below.

§2.4 <u>OTHER DOMAINS</u>. <u>Electromagnetism</u> leads to problems similar to those in acoustics concerning wave propagation across a dioptra : in linear electromagnetism, i. e. when the constitutive equations are $\vec{D} = \varepsilon \vec{E}$, $\vec{B} = \mu \vec{H}$, $\vec{J} = \gamma \vec{E}$ ($\vec{D}$ = electric induction (vector)), $\vec{E}$ = electric field, $\vec{B}$ = magnetic induction, $\vec{H}$ = magnetic field, $\vec{j}$ = electric current ; ε, μ and γ are respectively the dielectric constant, the magnetic permeability constant, the conductivity constant and depend on the medium, so that they are discontinuous on a dioptra) one has a system of linear partial differential equations (the Maxwell equations) with discontinuous coefficients, thus showing multiplications of distributions on the dioptra. In certain cases (for instance in the study of plasmas) physicists use nonlinear constitutive equations, see Serre[1] ; also in this case formal calculations work well (Serre [1] p232-233). Bampi-Zordan [1] suggest to model the electromagnetic field generated by a point charge having a jump in velocity by a system of conservation laws (5) with a solution that has not only a discontinuity but also a Dirac function on this discontinuity, thus showing products of distributions such as δ^2, $Y\delta$,...

The system of _hydrodynamics_ (10) is a system of conservation laws (5). But in case of chemical reactions and radiation losses one may be forced to use a nonconservative equation for the internal energy in order to obtain the temperature as correctly as possible (Hain [1]), thus showing products of distributions in case of shock waves. See also Bedeaux - Albano - Mazur [1] in thermodynamics. In the sequel of this book we shall also study a nonconservative version of the system of hydrodynamics, used for the numerical simulation of strong collisions. Setting $v = \dfrac{1}{\rho}$ (v is the specific volume) this system has the form

$$(17)\begin{cases} v_t + uv_x - vu_x = 0 \\ u_t + uu_x + vp_x = 0 \\ p_t + up_x + f(p,v)\, u_x = 0 \end{cases}$$

for usual equations of state (from which the function f follows). See also Lodder [1] in plasmaphysics, Raju [1,2] in relativity and astrophysics.

<u>Research problems.</u> There are as many research problems as specific situations ; they can consist in <u>theoretical mathematical works</u> (proofs of existence, uniqueness, justification of formal calculations) in <u>theoretical physical works</u> (when the usual "formal" statement of the equations appears to be ambiguous then one should state them more precisely (in $\mathcal{G}$) so as to resolve the ambiguity), see § 4.1 and § 4.2 below, or in <u>numerical works</u> (usually inspired by the two above works). Various examples will be given in the sequel of this book.

Chapter 3 . Elementary introduction

§3.1 AN INTUITIVE WAY TO CONCEIVE THESE "NEW GENERALIZED FUNCTIONS"

The goal of this heuristic section is to give to the reader an intuitive understanding of these "generalized functions". This is all the more so important as readers are expected to drop the precise mathematical definitions, and so they will work only with the presentation given in this section.

Shock waves are "very quick" variations of the physical variables that describe a given situation. But these variations are not "instantaneous" : they require a few mean free paths, i. e. the jumps involve several times the average distance between molecules. The situation can be depicted by Fig. 1 below

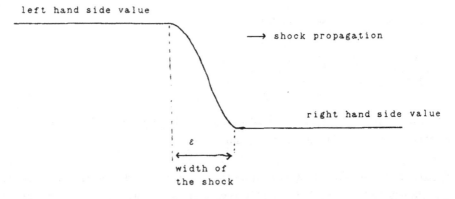

Figure 1. Propagation of a shock wave : realistic physical viewpoint ; to fix ideas $\varepsilon \# 10^{-8}$ meters for supersonic bangs in the air at the level of sea and in usual conditions.

Mathematically one represents the phenomenon by passing to the limit $\varepsilon \to 0$. Then the shock is represented by a discontinuous function : Fig 2 below

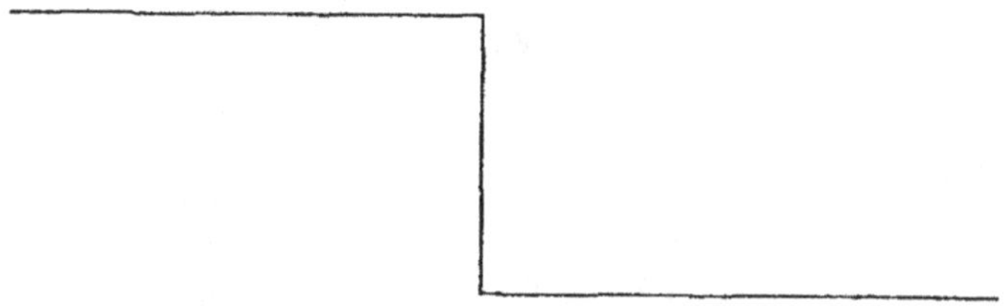

Figure 2. Mathematical description of the same shock wave.

In classical analysis one has no piece of information on the jump of a discontinuous function, as the one represented in Fig 2. The "physical" representation as in Fig 1 contains, at least in priciple (since measurements within the width of the shock wave are usually impossible), some information on the behavior of the physical variables within the shock. This information is lost in the passage from Fig 1 to Fig 2, i.e. in the usual mathematical idealization of the shock wave. In many cases this loss of information does not cause any trouble. But, especially in the cases considered in this book, it is at the very basis of the encountered difficulties. When one represents shock waves by generalized functions in our sense then there lies in the generalized functions themselves some piece of information on the jump, i.e. on the passage from the left hand side value to the right hand side value, quite similar to the situation depicted in Fig 1.

Let us show (repeating a calculation in §1.2) on "formally easy" calculations how the classical concept of a discontinuous function can lead to absurdities. Let Y be the Heaviside function as defined classically above in figure 2. For any integer n one has in the classical sense

(1) $$Y^n = Y.$$

Differentiation of (1) gives

(2) $$nY^{n-1}Y' = Y',$$

in particular

(2') $$YY' = \frac{1}{2} Y'.$$

Multiplication of both sides of (2') by Y gives

$$Y^2Y' = \frac{1}{2}YY'.$$

Then (2) and (2') give

$$\frac{1}{3}Y' = \frac{1}{4}Y'$$

which is absurd since $Y' = \delta$ is nonzero. In our theory we interpret this absurdity as a proof of the fact that (1) is false for $n \neq 1$. Indeed, if we consider that Y has some "continuous jump" or some "pointvalue" at the origin, then it is clear that Y and Y^n differ by these concepts : Fig 3.

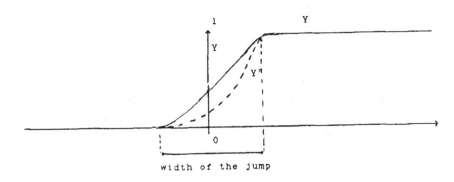

width of the jump

Figure 3. An intuitive depiction of a difference between Y and Y^n for $n > 1$, due to different "behaviors on the jump". When the "width of the jump" tends to zero there subsists something like "Y (0) $\neq Y^n$ (0)" which makes some difference between Y and Y^n.

Now let us consider the analogy between our construction of generalized functions and the construction of the real numbers. A real number r may be considered as some "ideal limit" of a sequence $\{r_n\}_{n \in \mathbb{N}}$ of decimal numbers, the decimal number r_n being the approximate value, up to the nth decimal, of the number r. The "exact value r" does not really make sense in physics. It makes sense mathematically but this is an "abstract trick". Let Ω be any open set in $\mathbb{R}^n$, $n = 1,2,...$. A generalized function G on Ω, in our sense, is an "ideal limit", when $\varepsilon \to 0$, of a family $\{R_\varepsilon\}_{0 < \varepsilon < 1}$ of C^∞ functions on Ω. The family $\{R_\varepsilon\}_{0 < \varepsilon < 1}$ is called a representative of G. Provided the differences $R_\varepsilon - Q_\varepsilon$ of functions in such two families of C^∞ functions would be "small enough" when $\varepsilon \to 0$ - in some sense to be made precise later - then the families $\{R_\varepsilon\}_{0 < \varepsilon < 1}$ and $\{Q_\varepsilon\}_{0 < \varepsilon < 1}$ represent the same generalized function. In particular if $R_\varepsilon = Q_\varepsilon$ (equality of C^∞ functions on Ω) as soon as $\varepsilon > 0$ is small enough then they represent the same generalized function. The converse does not hold : we may have $R_\varepsilon \neq Q_\varepsilon$ for all ε, at the same time as $\{R_\varepsilon\}_{0 < \varepsilon < 1}$ and $\{Q_\varepsilon\}_{0 < \varepsilon < 1}$ may represent the same generalized function (according to the definitions given in chapter 8 ; it holds with the simpler definitions given in Egorov [1]). Operations on generalized functions make sense when the results of the same operations on representatives do not depend, in the sense of their "ideal limit", on the arbitrariness in the choice of representatives.

These considerations are sufficient so that one could already perceive that products appearing in the form Y. δ are ambiguous. Let H be a fixed generalized function (in our sense) representing the Heaviside step function and let δ_i, $1 \leq i \leq 3$, be three generalized functions representing the Dirac delta function. Imagine that, for all $\varepsilon > 0$ small enough, their relative aspects are depicted by Fig 4 below :

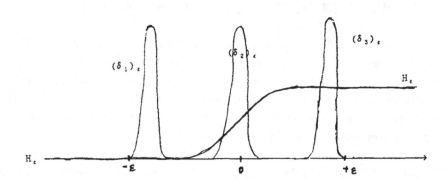

Figure 4. Three possibilities for the product H. δ.

Then it is quite clear that in this situation we have $H\delta_1 = 0$, $H\delta_3 = \delta_3$ and for instance something like $H\delta_2 = A\delta$ with $0 < A < 1$ and δ another Dirac like function. This reflects the fact that the classical product" Y. δ is ambiguous. One gets rid of this ambiguity, in each precise circumstance, by a suitable analysis of the relative "microscopic situation" of the generalized functions representing Y and δ in this circumstance. Of course the word "microscopic" is intended for the behavior of representatives $\{Y_\varepsilon\}_{0<\varepsilon<1}$ and $\{\delta_\varepsilon\}_{0<\varepsilon<1}$, of Y and δ respectively, for $\varepsilon > 0$ small enough i.e. arbitrarily close to the origin. In figure 4 there are "microscopic translations" between the supports of the δ's and the place in which the jump of H takes place : this relative situation influences completely the result. For instance if the δ under consideration is exactly the derivative of H (for instance if $\delta_\varepsilon = H_\varepsilon'$ for all $\varepsilon > 0$ small enough)then we are in a situation similar to the δ_2 case).

In conclusion of this section the idea in our approach is that the genuine physical objects should be represented by some C^∞ function R_ε, for $\varepsilon > 0$ unknown. Not only we do not usually know accurately this function R_ϵ but also it might be that this knowledge - theoretically possible at least in some ideal problems - would be useless - since too complicated. To formalize this ignorance we have to consider mathematical "ideal objects" which, for the applications developed in the sequel, are more subtle than the naive limit $\epsilon \to 0$ as considered in classical analysis. These objects are called "new generalized functions". To compute on them we may use general rules of computation - to be exposed in the next section - which follow from similar computations on representatives. In some cases the limit $\epsilon \to 0$ is not possible in intermediate calculations. Often this reflects the fact that these calculations do not make sense within classical analysis. In this event this theory gives the possibility to compute numbers that emerge from physical experiments, in cases in which this was previously impossible. This aspect will be developed in the sequel of this book.

§3.2 - DESCRIPTION OF THE RULES OF CALCULATION IN $\mathcal{G}(\Omega)$

We denote by $\mathcal{G}(\Omega)$ the set of all "new" generalized functions on Ω, Ω open set in $\mathbb{R}^n$. These generalized functions may be, according to the situation, real or complex valued. In all the physical situations encountered in this book they are assumed to be real valued ; anyway this cannot cause any trouble. An element of $\mathcal{G}(\Omega)$ is usually denoted by a capital letter such as G ; it may also be denoted by G (x) in case this notation could not bring confusion with its value G(x) at the point x (same possibility of confusion as for classical functions).

a) ALGEBRAIC OPERATIONS. We denote by $\mathcal{C}^\infty(\Omega)$ the set of all the classical C^∞ functions on the open set Ω (real valued or complex valued). We have a natural inclusion of $\mathcal{C}^\infty(\Omega)$ into $\mathcal{G}(\Omega)$ that we note

$$\mathcal{C}^\infty(\Omega) \subset \mathcal{G}(\Omega).$$

In $\mathcal{C}^\infty(\Omega)$ we have three classical operations : the addition, the multiplication of a function by a scalar and the multiplication of functions : all these operations extend to $\mathcal{G}(\Omega)$ with exactly all the same properties. We do not list these properties that everybody knows very well (for instance $(G_1 + G_2)G_3 = G_1 G_3 + G_2 G_3,...$). These operations are denoted in $\mathcal{G}(\Omega)$ by the same notations as the classical ones.

REMARK -From §1.3, §1.4 we know that the multiplication of distributions does not make sense in general, according to distribution theory. Since the multiplication of generalized functions always makes sense, and still gives an element of $\mathcal{G}(\Omega)$, this puts in evidence a basic difference between our theory and distribution theory.

b) RESTRICTIONS AND LOCAL PROPERTIES. If Ω' is an open subset of Ω and if $G \in \mathcal{G}(\Omega)$ then its restriction $G|\Omega'$ may be naturally defined as an element of $\mathcal{G}(\Omega')$, generalizing exactly the classical concept of restriction of C^∞ functions. Let $\{\Omega_i\}_{i \in I}$ be a family of open subsets of Ω (the set I plays the role of a set of indices, it is not necessarily a finite set) and let $\Omega' = \underset{i \in I}{U} \; \Omega_i$.

If $G \in \mathcal{G}(\Omega)$ is such that

$$\forall i \in I \quad G|\Omega_i = 0 \text{ in } \mathcal{G}(\Omega_i)$$

then

$$G|\Omega' = 0 \text{ in } \mathcal{G}(\Omega').$$

In particular any generalized function $G \in \mathcal{G}(\Omega)$ whose restrictions to a neighborhood of each point of Ω are null is the generalized function 0 (which is of course a classical C^∞ function). In short, exactly like the C^∞ functions, the generalized functions have a local character. We may define the support of a generalized function G as the complement of the union of all the open subsets in which the restrictions of G are null (exactly like in the classical case).

If $G \in \mathcal{G}(\Omega)$ and if L is a vector subspace of $\mathbb{R}^n$ such that $L \cap \Omega$ is nonvoid then the restriction $G|L \cap \Omega$ is defined as an element of $\mathcal{G}(L \cap \Omega)$.

Example : if $(x,t) \to G(x,t)$ is a generalized function on $\mathbb{R}^n \times \mathbb{R}$ ($x \in \mathbb{R}^n$, $t \in \mathbb{R}$) then the restriction $G|\mathbb{R}^n \times \{0\}$, i. e. the generalized function $(x,0) \to G(x,0)$, is naturally defined and is an element of $\mathcal{G}(\mathbb{R}^n)$. This concept is used for defining initial conditions for Cauchy problems.

c) PARTIAL DERIVATIVES. Let $D = \dfrac{\partial^{|k|}}{\partial x_1^{k_1}...\partial x_n^{k_n}}$ where $k_1,...,k_n$ are integers and where we set as usual $|k| = k_1 + ... + k_n$) be any partial derivation. Then for any $G \in \mathcal{G}(\Omega)$ its partial derivative DG is naturally defined as an element of $\mathcal{G}(\Omega)$. When G is a classical C^∞ function this concept of partial derivative coincides exactly with the usual one. Leibniz's rule $((uv)' = u'v + uv'$ and extensions for higher order derivatives and functions of several real variables) holds in $\mathcal{G}(\Omega)$.

d) More generally than products, various nonlinear functions of generalized functions (for instance sin G, cos G if $G \in \mathcal{G}(\Omega)$ is real valued) are naturally defined as elements of $\mathcal{G}(\Omega)$ and generalize exactly the corresponding classical operations on $\mathcal{C}^\infty$ functions.

In short , for all these operations we treat the generalized functions exactly in the same way as the classical $\mathscr{C}^\infty$ functions.

e) INTEGRATION THEORY, POINTVALUES AND GENERALIZED NUMBERS

We have an integration theory of generalized functions : let $G \in \mathscr{G}(\Omega)$ and let K be a closed bounded set in Ω (a compact subset of Ω). Then we may define the concept of the integral of G on K, denoted by

$$\int_K G(x) \, dx.$$

However, this integral is not in general a real or complex number : to define it in full generality we need to define a concept of generalized (real or complex) numbers. These generalized numbers may be infinitely small (in absolute value and at the same time nonzero). In this book we do not introduce them for simplification since we can limit ourselves to the case this integral gives a classical number. When G has a compact support K in Ω we define

$$\int_\Omega G(x) \, dx$$

as $\int_{K'} G(x) \, dx$ if K' is any compact subset of Ω containing K in its interior.

In a similar way if G is any generalized function on Ω and x any point of Ω then the value G(x) of G at the point x is defined as a generalized number.

Integration theory and pointvalues of generalized functions cannot lead to confusion since one computes on the generalized real (or complex) numbers exactly like in the case of classical numbers.

All of the above operations are trivially obtained (from the definitions) by reproducing the same operations on representatives : if G_1, $G_2 \in \mathscr{G}(\Omega)$ have the families $\{R_{1,\varepsilon}\}_{0<\varepsilon<1}$ and $\{R_{2,\varepsilon}\}_{0<\varepsilon<1}$ as respective representatives then the family $\{R_{1,\varepsilon} + R_{2,\varepsilon}\}_{0<\varepsilon<1}$ is a representative of $G_1 + G_2$, the family $\{R_{1,\varepsilon} \cdot R_{2,\varepsilon}\}_{0<\varepsilon<1}$ is a representative of $G_1 G_2$, the family $\{DR_{1,\varepsilon}\}_{0<\varepsilon<1}$ is a representative of DG_1 (D any partial derivation operator). Integration theory and pointvalues work exactly in the same way since generalized numbers are defined exactly by the same process than generalized functions, as ideal limits of a family of classical numbers depending on the parameter ε.

If K is a compact subset of Ω, if $G \in \mathcal{G}(\Omega)$ and if the family $\{R_\varepsilon\}_{0<\varepsilon<1}$ is a representative of G then the integral

$$I = \int_K G(x)\, dx$$

is the "ideal limit", see remark below, of the classical integrals

$$I_\varepsilon = \int_K R_\varepsilon(x)\, dx$$

when $\varepsilon \to 0$; obvious extension to the integral over Ω of a generalized function with compact support in Ω.

EXAMPLE. Let us consider the various generalized functions δ_i considered in Fig. 4. Since by definition $\int_{-\infty}^{+\infty} (\delta_i)_\varepsilon(x)\, dx = 1$ one gets

$$\int_{-\infty}^{+\infty} (\delta_i)(x)\, dx = 1.$$

EXAMPLE. Let $\mathcal{D}(\Omega)$ denote the set of all C^∞ functions on Ω with compact support. For any $\Psi \in \mathcal{D}(\mathbb{R})$, what is

$$\int_{-\infty}^{+\infty} (\delta_i)(x)\, \Psi(x)\, dx \ ?$$

Answer : The limit $\varepsilon \to 0$ gives of course $\Psi(0)$; using the precise definitions in Colombeau [3] one obtains exactly $\Psi(0)$.

REMARK. The definition of the integral as a usual limit would be an abusive simplification of the more precise definition given in Colombeau [3] : it could lead to mistakes in calculations ; the correct statement is : I is a generalized number which is the class of $\{I_\varepsilon\}_{0<\varepsilon<1}$. When the limit of I_ε, for ε tending to 0, does not exist in $\mathbb{R}$ or $\mathbb{C}$, this more sophisticated definition applies. According to it

$$\int_K G(x)\, dx \text{ is a "generalized number" :}$$

<u>EXAMPLE</u> : $\int \delta^2 (x)\, dx$ is "infinite" in some sense made precise in Colombeau [3].

If $G \in \mathcal{G}(\Omega)$ and $x \in \Omega$ then we naturally define the value $G(x)$ of G at the point x as the limit when it exists of $R_\varepsilon (x)$ when $\varepsilon \to 0$. Exactly the same remarks as above should be done concerning pointvalues ; for instance the pointvalue $\delta(0)$, δ a Dirac delta function, is intuitively infinite; it makes sense as a generalized number, not as a classical number (in fact in our theory there are several "Dirac generalized functions" and their value at the origin can depend on the one under consideration).

f) A NOVELTY : THE WEAK EQUALITY OR "ASSOCIATION". A basic role is played by a relation that can hold between two different elements of $\mathcal{G}(\Omega)$: $G_1, G_2 \in \mathcal{G}(\Omega)$ are said to be <u>associated</u> iff for any $\psi \in \mathcal{D}(\Omega)$ the integral

$$\int_\Omega (G_1(x) - G_2 (x))\, \psi(x)\, dx$$

gives as a result the classical number 0 in the sense above (of the usual limit of integrals I_ε when $\varepsilon \to 0$). This means that if $\{R_{1,\varepsilon}\}_{0<\varepsilon<1}$ and $\{R_{2,\varepsilon}\}_{0<\varepsilon<1}$ are respective representatives of G_1 and G_2 then for any $\Psi \in \mathcal{D}(\Omega)$ the classical integral

$$I_\varepsilon = \int_\Omega (R_{1,\varepsilon}(x) - R_{2,\varepsilon} (x))\, \psi (x)\, dx$$

tends to 0 when $\varepsilon \to 0$. It follows from the definitions that if the above holds for a particular pair of representatives then it also holds for any pair of respective representatives.

Two associated generalized functions are not equal in general. This will follow from the precise definitions. Indeed let us consider the generalized functions Y^n and Y in Fig. 3. Then they are associated and it has been proved there that they are different elements of $\mathcal{G}(\mathbb{R})$.

<u>Notation</u> : a new notation is requested for the association ; we write $G_1 \approx G_2$. Both members of an association may be differentiated at will :

$$G_1 \approx G_2 \Rightarrow DG_1 \approx DG_2$$

for any partial differential operator D. It is obvious that the association is an equivalence relation. This relation is compatible with the addition ($G_1 \approx G_2$, $G_3 \approx G_4 \Rightarrow G_1 + G_3 \approx G_2 + G_4$), with the differentiation (above) and with scalar multiplication. However, the association is not compatible with the multiplication, i. e. in general $G_1 \approx G_2$ does not imply $GG_1 \approx GG_2$ for $G, G_1, G_2 \in \mathcal{G}(\Omega)$. Even $G \approx 0$ does not imply $G^2 \approx 0$ in general. As an example let ρ be a given positive C^∞ function on $\mathbb{R}$, for which $\rho^{1/2}$ is also C^∞, null outside $[-1, +1]$ and such that

$$\int \rho(\lambda) \, d\lambda = 1.$$ Let G be defined as the class of $R_\varepsilon(x) = (\varepsilon)^{-1/2} \rho^{1/2} (\frac{x}{\varepsilon})$. Then it is immediate to

check that G^2 has the properties of the Dirac delta function while G is associated with 0. A similar concept of association can be defined in the set of our generalized numbers.

Remark . As it might be guessed easily from the definitions of association and of a distribution the concept of association extends the classical concept of equality of distributions. Due to the fact that the symbol = was already needed for the equality in $\mathcal{G}(\Omega)$ (i. e. the equality of two elements of a set), we had to introduce a new symbol for the association. The reader must be aware of the fact that very often it is the concept of association which represents faithfully the classical concept of equality of integrable functions or distributions. The fact that the association is not coherent with the multiplication reflects the well known fact that the multiplication of distributions is in general impossible within classical analysis. In short the classical concept of equality of functions or distributions splits into two concepts (= and $\approx$) that should be used in their right place in the applications.

TERMINOLOGY. Due to an obvious interpretation in physical cases we often say that G_1 and G_2 have the *same macroscopic aspect* to express that they are associated. In case $G_1 \neq G_2$ and $G_1 \approx G_2$ we say they do not have the *same microscopic aspect* although they have the same macroscopic aspect.

§3.3 - EXAMPLES OF CALCULATIONS

3.3.1. Heaviside generalized functions. We have already shown that in $\mathcal{G}(\mathbb{R})$ there are several elements H which have the property of the classical Heaviside function. We call them "Heaviside generalized functions". More precisely :

Definition. Let $H \in \mathcal{G}(\mathbb{R})$ with representative $\{R_\varepsilon\}_{0<\varepsilon<1}$.
We say that H is an "Heaviside generalized function" iff

a) the restriction of H to $(-\infty, 0[$ is the zero function
b) the restriction of H to $]0, +\infty)$ is the function identical to one
c) $\underset{\substack{0<\varepsilon<\eta \\ -1 \le x \le 1}}{\operatorname{Sup}} |R_\varepsilon(x)| < +\infty$ for some $\eta > 0$.

As a consequence

$$\lim_{\varepsilon \to 0} \int R_\varepsilon (x) \, \psi (x) \, dx = \int_0^{+\infty} \psi (x) \, dx$$

for any $\psi \in \mathcal{D}(\mathbb{R})$. Therefore all Heaviside generalized functions are associated with each other. Let $n = 2,3,...$ then H^n is a Heaviside generalized function as soon as H is. Thus

$$H^n \approx H.$$

Differentiation gives the formula

(3)
$$H^n H' \approx \frac{1}{n+1} H',$$

in particular

(3')
$$HH' \approx \frac{1}{2} H'.$$

Remark. More general definitions are possible by replacing c) by some weaker assumption ; the above one is convenient and sufficiently general for the applications in this book.

3.3.2. Discontinuous solutions of the equation $u_t + uu_x = 0$.
One of the simplest models of nonlinear partial differential equations is the inviscid Burgers equation

(4)
$$u_t + uu_x = 0 \quad \text{(strong equality in } \mathcal{G}(\mathbb{R}^2))$$

that we can also state in the weak form

(5)
$$u_t + uu_x \approx 0 \quad \text{(association in } \mathcal{G}(\mathbb{R}^2)).$$

Let us seek steady shocks (also called travelling wave solutions)

(6)
$$u(x,t) = \Delta u \, H(x-ct) + u_\ell$$

where u_r, u_ℓ are respectively the right and left hand side values (assumed to be constant), $\Delta u = u_r - u_\ell$ is the jump , c = constant is the velocity of the shock ; H is a Heaviside generalized function.

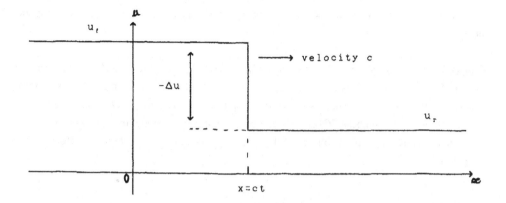

Figure 5. The steady shock represented by (6).

Putting (6) into (5) :

$$-c\Delta uH' + (\Delta uH + u_\ell)\,\Delta uH' \approx 0.$$

From (3') and after division by Δu (which is assumed nonzero)

$$c - u_\ell = \frac{\Delta u}{2}$$

i.e

(7)
$$c = \frac{u_\ell + u_r}{2}.$$

This is the well known jump condition for the inviscid Burgers' equation. Indeed this calculation proves that (6) is solution of (5) if and only if (7) holds. Now let us notice that any solution u of (4) is necessarily also solution of the equation $uu_t + u^2u_x = 0$ since the equality in $\mathcal{G}(\mathbb{R}^2)$ is compatible with the multiplication. This last equation can be equivalently stated as $\frac{1}{2}(u^2)_t + \frac{1}{3}(u^3)_x = 0$. Putting (6) into any of these two equivalent formulations one gets at once the jump condition

(8)
$$c = \frac{2}{3}\frac{(u_\ell)^2 + (u_r)^2 + u_\ell u_r}{u_\ell + u_r}.$$

(7) and (8) are contradictory ! This proves that indeed the strong formulation (4) has no solution of the form (6) (when $\Delta u \neq 0$). Note that since the association is incompatible with the multiplication (5) does not imply $uu_t + u^2 u_x \approx 0$, and so one cannot obtain this contradiction from the weak formulation (5). We have proved in this example that the statements of the equations with the equality and with the association usually lead to quite different results in the case of shock waves : the statement with the equality in $\mathcal{G}$ often gives no discontinuous solutions ; the statement with the

association is closely related with the classical concept of a weak solution (in the sense of distribution theory).

Remark In the case of systems of conservation laws (formula 5 in chap. 2) and in the presence of uniform L^∞ (or L^p) estimates Di Perna has considered a concept of "measure valued solutions", weaker than the concept of weak (= distributional) solution, see Di Perna [1], Di Perna - Majda [1]. It has been proved in Colombeau - Oberguggenberger [3] that this concept of measure valued solutions is equivalent to our concept of solution in the sense of association, i. e. statement of (formula 5 in chap. 2) as

$$\vec{U}_t \approx \text{Div } f(\vec{U}).$$

3.3.3. A system in nonconservative form. We consider the system of two equations

(9)
$$\begin{cases} u_t + uu_x \approx \sigma_x \\ \sigma_t + u\sigma_x \approx k^2 u_x \end{cases}$$

where $k > 0$ is a constant, and where the two unknown functions $u(x,t)$ and $\sigma(x,t)$ have to be sought in $\mathcal{G}(\mathbb{R}^2)$. (9) is in nonconservative form (definition in § 2.1) because of the term $u\sigma_x$. Now let us seek travelling waves (i.e. represented by fig. 5) solutions of (9). We set

(10)
$$\begin{cases} u(x,t) = \Delta u \, H(x - ct) + u_\ell \\ \sigma(x,t) = \Delta\sigma \, K(x - ct) + \sigma_\ell \end{cases}$$

with Δu, $\Delta\sigma$, u_ℓ, σ_ℓ real numbers and H, K two Heaviside generalized functions.

3.3.4. Remark .There is a priori no reason to choose H = K ; from our assumption that u and σ are travelling waves we only know that H and K are Heaviside generalized functions, not necessarily the same.

3.3.5 Exercise. Prove that when (9) is stated with two equality symbols (in place of the two association symbols) in $\mathcal{G}(\mathbb{R}^2)$ then it has no solution of the form (10) (with $\Delta u \neq 0$, $\Delta\sigma \neq 0$). The proof is similar to that in 3.3.2 above ; this result will also follow from thm. 4.1.3 below.

3.3.6. The jump conditions for system (9). Putting (10) into (9) we get the relations

(11)
$$\begin{cases} (c - u_\ell) \Delta u H' - (\Delta u)^2 HH' + \Delta\sigma K' \approx 0 \\ (c - u_\ell) \Delta\sigma K' - \Delta u\Delta\sigma HK' + K^2 \Delta u H' \approx 0. \end{cases}$$

From (3') the first line of (11) gives ($\Delta u \neq 0$)

(11a)
$$c - u_\ell = \frac{\Delta u}{2} - \frac{\Delta \sigma}{\Delta u}$$

which is the classical Rankine - Hugoniot jump relation for the first line of (9). Since $\Delta u \Delta \sigma \neq 0$ the second line of (11) implies that

(11b)
$$HK' \approx A\delta$$

(δ a Dirac function, for instance $\delta = H'$ or $\delta = K'$, A a real number) and thus the second line of (11) gives

(11c)
$$c - u_\ell = A\Delta u - k^2 \frac{\Delta u}{\Delta \sigma} .$$

Conversely the relations (11a,b,c) imply that (10) is solution of (9). We have obtained

3.3.7. Theorem – System (9) admits an infinite number of jump conditions for shocks of the form (10) (travelling waves). These conditions are (11a, b, c) and depend on an arbitrary number A. This number A reflects the relative "microscopic shapes" of the jumps of H and K.

3.3.8. Remark This result and the above calculations are completely new since they are meaningless within distribution theory. This result shows a major difference between the nonconservative case (an infinity of possible jump relations) and the conservative case (a unique jump relation given by the classical Rankine Hugoniot jump formulas).

§3.4 NUMERICAL EVIDENCE OF AN INFINITE NUMBER OF JUMP CONDITIONS

In this section we put in evidence numerically theorem 3.3.7 on the existence of an infinite number of jump conditions for system (9). This is done as follows : one considers natural, slightly different schemes, which, in case of a smooth solution or in case of a system in conservation form, always lead to the same (classical) solution. Then, in the case of system (9), we observe that these different schemes converge to different solutions of (9), corresponding to different values of the parameter A in 11 a) c). From 11b) one can estimate numerically the value of A (from an observation of the relative aspects of the jumps of u and σ on the shocks : "microscopic profile of the shocks"). One finds agreement between the values of A computed from the step values using 11a) c) and between the values of A directly observed from the microscopic profile of the shocks.

3.4.1. Numerical schemes . Let us consider the following scheme ; $r = \dfrac{\Delta t}{h}$ where Δt and h are the respective time and space steps (usual index notations : u_i^n is the value of u in the mesh $(i - \frac{1}{2})h < x < (i + \frac{1}{2})h$, $t = n\Delta t$).

$$(S_0) \begin{cases} u_{i+1}^{n+1} = m_i^n - r\, m_i^n\, \dfrac{u_{i+1}^n - u_{i-1}^n}{2} + r\, \dfrac{\sigma_{i+1}^n - \sigma_{i-1}^n}{2} \\[3mm] \sigma_{i+1}^{n+1} = g_i^n - rm_i^n\, \dfrac{\sigma_{i+1}^n - \sigma_{i-1}^n}{2} + rk^2\, \dfrac{u_{i+1}^n - u_{i-1}^n}{2} \\[3mm] m_i^n = \dfrac{u_{i+1}^n + 2u_i^n + u_{i-1}^n}{4} \;,\; g_i^n = \dfrac{\sigma_{i+1}^n + 2\sigma_i^n + \sigma_{i-1}^n}{4} \end{cases}.$$

As this is classical the mean values m_i^n and g_i^n are introduced to ensure stability ; of course they amount to a viscosity. One can prove that this scheme is stable for the L^∞ norm and preserves the bounded variation in space as well as the bounded variation in time in the sense of Tonelli - Cesari ; then one can deduce from these properties that this scheme converges to a solution of (9) in our sense (see Cauret - Colombeau - Le Roux [1], and Biagioni [1] chap 3 for a proof in a similar case and Noussair [1] for a proof in the case of (S_0)). Now let us consider, for $\lambda \in \mathbb{Z}$, the schemes (S_λ) obtained from (S_0) by translating u by $|\lambda|$ meshes, to the left or to the right side :

$$(S_\lambda) \begin{cases} u_i^n = v_{i+\lambda}^n \\[6pt] v_{i+1}^{n+1} = l_i^n - rl_i^n(v_{i+1}^n - v_{i-1}^n)/2 + r(\sigma_{i+1}^n - \sigma_{i-1}^n)/2 \\[6pt] \sigma_{i+1}^{n+1} = g_i^n - rm_i^n(\sigma_{i+1}^n - \sigma_{i-1}^n)/2 + rk^2(v_{i+1}^n - v_{i-1}^n)/2 \\[6pt] l_i^n = \dfrac{v_{i+1}^n + 2v_i^n + v_{i-1}^n}{4}, \; m_i^n = \dfrac{u_{i+1}^n + 2u_i^n + u_{i-1}^n}{4}, \\[12pt] g_i^n = \dfrac{\sigma_{i+1}^n + 2\sigma_i^n + \sigma_{i-1}^n}{4} \end{cases} .$$

In the case of smooth solutions, of course (S_λ) converges to the same solutions as (S_0). We have not been able to prove for (S_λ) BV properties. One can prove, exactly as in Cauret - Colombeau - Le Roux [1], Biagioni [1] chap 3, that, assuming (S_λ) has the BV properties, then it would still converge to a solution of (9), see Noussair [1].

Figure (6) below shows that the three schemes corresponding to the values $\lambda = -3$, $\lambda = 0$ and $\lambda = 3$ give very different solutions. One observes that (S_0) corresponds exactly to A = 1/2 while (S_{-3}) corresponds to values of A close to 0 and (S_3) to values of A close to 1 (Fig. 7,8,9).

3.4.2. The microscopic profile of shocks. As already explained there are several Heaviside functions in $\mathcal{G}(\mathbb{R})$; If H = K then A = 1/2. Thus the Heaviside functions H and K in the shocks of (S_{-3}), (S_0) or (S_3) are very different. The pair (H,K) for a given shock is called the "microscopic profile " of this shock. The quantity A in 11b) is a consequence of this microscopic profile. Thus we have checked on numerical tests that the microscopic profile of shocks rules the jump conditions (and the velocity) in case of nonconservative shocks.

In contrast, for conservative shocks, the jump conditions are independent of this microscopic profile (Rankine - Hugoniot formulas). This microscopic profile makes sense mathematically in the setting of $\mathcal{G}$ since there are several Heaviside generalized functions in $\mathcal{G}$; it does not make sense in classical analysis, in which this rather subtle distinction between the behavior of Heaviside functions on their point of discontinuity is ignored. This is precisely the reason for which discontinuous solutions of nonconservative systems do not make sense in $\mathcal{D}'$: the crucial fact ruling them cannot be expressed within $\mathcal{D}'$.

3.4.3 Remark. The profile of a shock can be observed as soon as the shock involves a few (4,5...) meshes. Antidiffusion techniques can be used provided they preserve this fact ; indeed a classical antidiffusion technique is used in the figures 6,7,8,9 in order to make the shocks neater : a correction is introduced in the two cases

(a)
$$u_{i-2}^n > u_{i-1}^n > u_i^n > u_{i+1}^n$$

(b)
$$u_{i-2}^n < u_{i-1}^n < u_i^n < u_{i+1}^n.$$

We set

$$c_i^n = \pm \min \left(\frac{|u_i^n - u_{i-1}^n|}{4}, \frac{|u_{i-2}^n - u_{i-1}^n|}{2}, \frac{|u_{i+1}^n - u_i^n|}{2} \right)$$

with + in the case (b) and - in the case (a) ; then u_i^n is replaced by $u_i^n + c_i^n - c_{i-1}^n$. Similar correction for σ_i^n.

3.4.4 Remark. The microscopic profile of H and K is apparent from the curves below. We focus our attention on the number A : its value as computed by 11 a) c) from the knowledge of the value of the steps is given for each shock in fig. 7,8,9. At the same time the value of A appears as the area of the darkened surface limited by the curve HK' and the horizontal axis when the area of the larger surface limited by the curve H' and the horizontal axis is taken to be one. One observes a complete agreement.

In fig. 6 $u_\ell = 3$ and in fig. 7,8,9 $u_\ell = 1$; in all cases $u_r = 0 = \sigma_\ell = \sigma_r$; $k^2 = 1$.

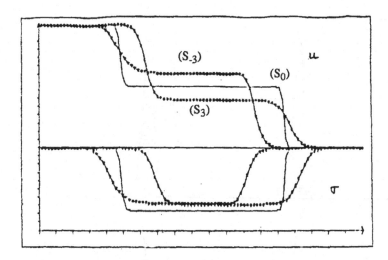

Figure 6. Superposition of the numerical solutions : an evidence of the fact there is an infinity of solutions.

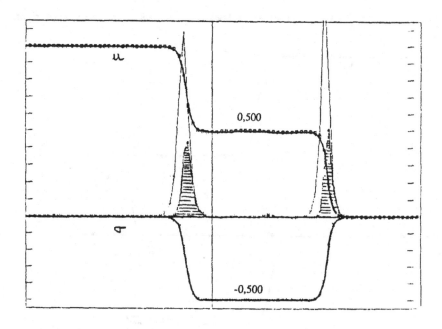

Figure 7. Case $\lambda = 0$; one finds A = 0,5 for both shocks

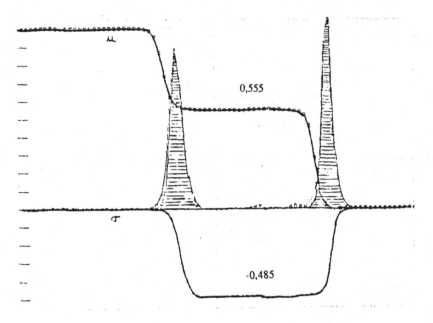

Figure 8. Case λ = 3 ; one finds A = 0, 89 for the left shock and A = 0, 98 for the shock on the right.

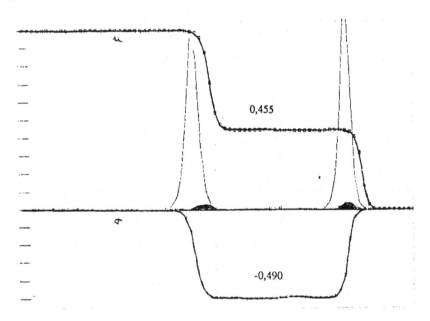

Figure 9. Case λ = −3 ; one finds A = 0, 05 for the left shock and A = 0, 06 for the shock on the right.

§.3.5 PROBLEM SECTION.

Problem 3.5.1. Consider the system

$$(12) \quad \begin{cases} v_t - u_x \approx 0 \\ u_t + (p(v))_x \approx 0 \end{cases}$$

with p a $\mathcal{C}^\infty$ function.

Question 1. Seek the jump conditions. State the system of four algebraic equations in the four unknowns $c_1, c_2, \bar{u}, \bar{v}$ corresponding to the solution of the Riemann problem with (u_g, v_g) and (u_d, v_d) as given values of u,v for x < 0 and x > 0 respectively (in case of two shock waves of velocities c_1 and c_2, $c_2 > c_1$, separated by a constant state $(\bar{u}, \bar{v})$; conventional representation :

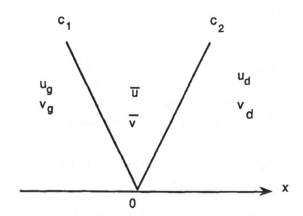

Solve the equations in the case $v_g = v_d = u_d = 0$, $u_g > 0$, $p(v) = v^2$.
What about the case $u_g < 0$ (with other same values)?

Question 2. a) Characterize the travelling wave solutions of the system

$$(12a) \quad \begin{cases} v_t - u_x = 0 \\ u_t + (p(v))_x \approx 0. \end{cases}$$

b) Characterize the travelling wave solutions of the system

$$(12b) \quad \begin{cases} v_t - u_x \approx 0 \\ u_t + (p(v))_x = 0. \end{cases}$$

Question 3. Same questions for the system

$$(12c) \quad \begin{cases} v_t - u_x = 0 \\ u_t + (p(v))_x = 0 \end{cases}.$$

in the cases

      a) $p(v) = v$

      b) $p(v) = -v$

      c) $p(v) = v^2$.

Sketch of answers 2a) same Heaviside generalized functions for u and v (plus the Rankine Hugoniot conditions). 2.b) a relation between the Heaviside generalized functions (plus the R. H. conditions). 3 a) no solution, 3. b) one solution, 3. c) no solution.

Problem 3.5.2. Consider the system

$$(13) \quad \begin{cases} u_t + (u^2)_x \approx q_x \\ q_t + (\frac{1}{3} u^3)_x \approx k^2 u_x, \ k > 0 \text{ constant.} \end{cases}$$

Question 1. Give the jump conditions ; if c denotes the velocity of the travelling wave prove the relation $(-c + u_\ell)^2 + (-c + u_\ell) \Delta u + \frac{(\Delta u)^2}{3} - k^2 = 0$; what happens in case $(\Delta u)^2 > 12k^2$? Counting the number of equations and of unknowns leads to the same form of the solution of the Riemann problem as in Ex 3. 5. 1 (two shock waves separated by a constant state). Is this possible when $|u_d - u_g| > 4k \sqrt{3}$?

Question 2. For $|u_d - u_g| > 4k\sqrt{3}$ one observes numerically strange phenomena (see the results of the Lax-Friedrichs 's scheme and of the Lax-Wendroff scheme : figure 10).

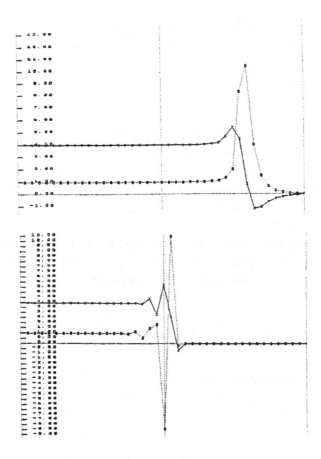

<u>Figure 10</u>. Numerical tests on system (13) in the case of question 2: the left and right step values are the given initial data of the Riemann problem : thus the solution of the Riemann problem appears to be made of only one "singular" shock.

One might try to understand them as follows by considering "singular travelling waves" of the form :

$$(ST) \quad \begin{cases} u(x,t) = \Delta u \, F(x - ct) + u_\ell & \text{with } u_r = u_d, \, u_\ell = u_g \\ q(x,t) = \Delta q \, G(x - ct) + q_\ell & \text{with } q_r = q_d, \, q_\ell = q_g \end{cases}$$

where $F, G, \in \mathcal{G}(\mathbb{R})$ are such that

$$F(\xi) = 0 = G(\xi) \qquad \text{if } \xi < 0$$
$$F(\xi) = 1 = G(\xi) \qquad \text{if } \xi > 0$$

(F and G are not necessarily Heaviside generalized functions since arbitrary singularities at $x = 0$ are allowed). Show that this attempt does not work.

<u>Question 3.</u> Let ε be a constant generalized function nonzero, but infinitesimal, (i.e. "infinitely small in absolute value") (or equivalently an infinitesimal but non zero generalized real number : ε may be viewed as the class of the map $R(\varepsilon,x) = \varepsilon$, $0 < \varepsilon < 1$). Consider the viscous system

$$(V) \quad \begin{pmatrix} u \\ q \end{pmatrix}_t + \begin{pmatrix} \dfrac{u^2 - q}{\frac{1}{3}u^3 - k^2 u} \end{pmatrix}_x \approx \varepsilon t \begin{pmatrix} u \\ q \end{pmatrix}_{x,x}$$

Does it have solutions of the form (10) or (ST)

<u>Question 4.</u> Consider the system

$$(NC) \quad \begin{pmatrix} u \\ q \end{pmatrix}_t + \begin{pmatrix} \dfrac{u^2 - q}{\frac{1}{3}u^3 - k^2 u} \end{pmatrix}_x \approx \varepsilon \begin{pmatrix} uq_x \\ 0 \end{pmatrix}$$

Does it has solutions of the form (10) ? Do you see from this a possible interpretation of the observed phenomenon in the case F and G would not change with time ?

<u>Question 5.</u> Check that for every $\alpha, \beta \in \mathbb{R}$ such that

$$\frac{\beta^2}{\alpha} + \frac{\alpha^2}{12} \leq 1$$

there is a solution of (13) in the form

$$u(x,t) = \alpha \, H(x - ct) + \beta\sqrt{t} \, f(x - ct)$$
$$q(x,t) = (\alpha^2 - c\,\alpha) \, H(x - ct) + \beta^2 \, t \, f^2(x - ct)$$

where $c = \frac{\alpha}{2} \pm \sqrt{1 - \frac{\beta^2}{\alpha} + \frac{\alpha^2}{12}}$, H is a generalized Heaviside function, f is a generalized function of support at the origin and such that

$$f \approx 0,\ f^3 \approx 0,\ f^2 \approx \delta\ (\delta = \text{the Dirac function})$$
$$Hf \approx 0,\ H^2 f \approx 0,\ (\alpha H - c)\, f^2 \approx 0.$$

Comments : f appears as some square root of δ.

Do you see a possible interpretation of the observed phenomenon in case - due to the factors $\sqrt{t}$ and t above - the functions F and G like in (ST) change with time ? Do numerical tests.

Sketch of answers. 1) if $\Delta u^2 > 12k^2$ there are no real numbers c and u_ℓ satisfying the jump relations ; when $|u_d - u_g| > 4k \sqrt{3}$ there is no solution of the Riemann problem by shock waves. This fact which was noticed in Colombeau - Le Roux [1] as well as the numerical results in question 2 led to the rejection of the conservative system (c) as a model for elasticity in the case of large $|\Delta u|$.

2) singular travelling waves of the form (ST) lead (integration of the relations obtained by putting them into (c)) to the same jump relations as usual travelling waves.

3) one still gets - as above in question 2 - the same jump relations. Note that the system (V) has been studied in Keyfitz-Kranzer [1] where solutions similar to those above are exhibited.

4) one gets $HK' \approx A\delta$, where A depending on ε is a generalized real number ; thus one has two unknowns : the velocity c and the generalized number A, for two equations. A can be observed "on the screen" as in §4. 4: when the two Heaviside functions H and K grow from 0 to 1, in a monotonic way then one has obviously $0 \le A\,(\varepsilon) \le 1\ \forall\ \varepsilon \in\ [0,1]$. Values of A $(\varepsilon) < 0$ or > 1 imply a "strange behavior" of H and K as those observed in fig 10. Thus our attempt of explanation : in the case $|u_d - u_g| > 4k\sqrt{3}$ the schemes would converge to solutions of a system with nonconservative second members, and so several quantities like A would raise the number of unknowns to at least two. These quantities would appear as peaks or oscillations in the jumps of the Heaviside functions. Note that a similar fact has been proved in Adamczewski - Colombeau - Le Roux [1] for a single equation.

5) This interpretation has been shown to the author by B. Keyfitz (private communication), see also Keyfitz-Kranzer [1].

Problem 3.5.3 Consider the system

$$\text{(S)} \qquad \begin{cases} u_t + uu_x \approx \sigma_x \\ \sigma_t + u\sigma_x \approx u_x \end{cases}$$

and assume that the solutions under consideration are represented by the same Heaviside generalized function on a shock, i. e.

$$\begin{cases} u(x,t) = \Delta u \; H(x - ct) + u_\ell \\ \sigma(x,t) = \Delta \sigma \; H(x - ct) + \sigma_\ell \end{cases}$$

<u>Question 1.</u> Give the jump conditions.

<u>Question 2.</u> Assume that the Riemann problem is to be solved by two successive shock waves (separated by constant states). Give the explicit formulas for $\bar{u}$, $\bar{\sigma}$, c_1, c_2 as functions of u_g, σ_g, u_d, σ_d in the case $0 < u_g - u_d < 4$; notation :

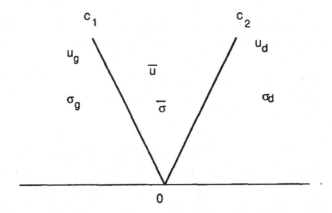

What is the sign of c_1 in the case $3u_g + u_d - \sigma_g + \sigma_d < 4$?

What is the sign of c_2 in the case $u_g + 3u_d - \sigma_g + \sigma_d > -4$?

In the sequel we shall always assume that $0 < u_g - u_d < 4$, $3u_g + u_d - \sigma_g + \sigma_d < 4$ and $u_g + 3u_d - \sigma_g + \sigma_d > -4$.

<u>Question 3.</u> Write a Godunov type scheme (see 4.6.3 below for a description of the Godunov scheme on a system of 3 equations) : this consists in choosing (in the projection step) for u_i^{n+1}, σ_i^{n+1} the mean value on the interval $(i - \frac{1}{2})h \leq x \leq (i + \frac{1}{2})h$ of the solution at time $(n+1)\,\Delta t$ of the Riemann problems stemming from the solution at time $n \, \Delta t$. The set of notations is (slightly different from the one adopted in question 2) :

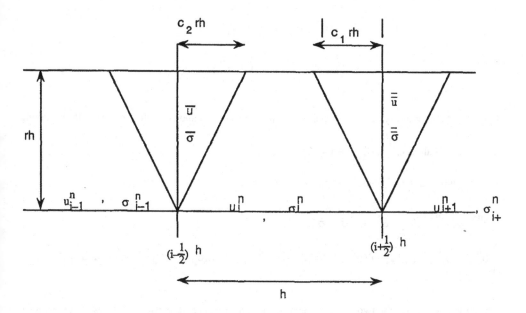

If $|u_i^n| \leq A \ \forall i \in \mathbb{Z}$ and $\forall n \in \mathbb{N}$ and if $|\sigma_{i+1}^n - \sigma_i^n| \leq B \ \forall i$ and $\forall n$ prove that this scheme can be

used as soon as $r < \dfrac{1}{2+2A+\dfrac{B}{2}}$ (stability condition).

<u>Answer</u> : one finds the set of relations

$$
\begin{cases}
u_i^{n+1} = c_2 r \ \bar{u} + (1 - c_2 r + c_1 r) u_i^n - c_1 r \ \bar{\bar{u}} \\[2mm]
\sigma_i^{n+1} = c_2 r \ \bar{\sigma} + (1 - c_2 r + c_1 r) \sigma_i^n - c_1 r \ \bar{\bar{\sigma}}
\end{cases}
$$

where

$$c_1 = -1 + \frac{1}{4}(3u_i^n + u_{i+1}^n - \sigma_i^n + \sigma_{i+1}^n)$$

$$c_2 = 1 + \frac{1}{4}(3u_i^n + u_{i-1}^n - \sigma_{i-1}^n + \sigma_i^n)$$

$$\bar{u} = \frac{1}{2}(u_{i-1}^n + u_i^n - \sigma_{i-1}^n + \sigma_i^n)$$

$$\bar{\sigma} = \frac{1}{2}(-u_{i-1}^n + u_i^n + \sigma_{i-1}^n + \sigma_i^n)$$

$$\bar{\bar{u}} = \frac{1}{2}(u_i^n + u_{i+1}^n - \sigma_i^n + \sigma_{i+1}^n)$$

$$\bar{\bar{\sigma}} = \frac{1}{2}(-u_i^n + u_{i+1}^n + \sigma_i^n + \sigma_{i+1}^n).$$

<u>Question 4.</u> Integrate system (S) in the cell $n \, \Delta t \leq t \leq (n+1) \, \Delta t$ and $(i - \frac{1}{2})h \leq x \leq (i + \frac{1}{2})h$, by taking as boundary values u_i^{n+1}, σ_i^{n+1} if $t = (n+1) \, \Delta t$.

<u>Answer,</u> with same notations as in question 3 one finds

$$\begin{cases} u_i^{n+1} = u_i^n - \frac{r}{2}(\bar{\bar{u}}^2 - \bar{u}^2) + r(\bar{\bar{\sigma}} - \bar{\sigma}) \\ \sigma_i^{n+1} = \sigma_i^n - \frac{r}{2}\{(u_i^n - \bar{u})(\sigma_i^n - \bar{\sigma}) + (\bar{u} - u_i^n)(\bar{\sigma} - \sigma_i^n)\} - r\{\bar{u}(\sigma_i^n - \bar{\sigma}) + u_i^n(\bar{\sigma} - \sigma_i^n) - (\bar{u} - \bar{\bar{u}})\}. \end{cases}$$

<u>Question 5.</u> Compare the two schemes obtained in questions 3 and question 4 : show that the values for u_i^{n+1} and σ_i^{n+1} are the same.

<u>Sketch of answers.</u> Question 1 is solved in 3.3.6 with $H = K$ i. e. $A = \frac{1}{2}$. For question 2 one has

$$\begin{cases} \bar{\sigma} - \sigma_g = \pm(\bar{u} - u_g) \\ \sigma_d - \bar{\sigma} = \pm(u_d - \bar{u}) \end{cases}$$

which gives four possibilities. The combinations $\overset{+}{_+}$ and $\overset{-}{_-}$ are at once eliminated since the values of $\sigma_g, \sigma_d, u_g, u_d$ are arbitrary. The condition $c_1 < c_2$ together with $0 < u_g - u_d < 4$ imposes the combination $\pm$. In the cases under consideration one has $c_1 < 0$ and $c_2 > 0$. The scheme in question 3 can be used provided the two shock waves inside the cell do not intersect. In question 4 the integration of the first equation is immediate since it is in conservation form. For the second equation one evaluates the quantity $u\sigma_x$ on each shock wave : for the one on the left one has

$$(u\sigma_x)(x,t) = (u_i^n - \bar{u})(\sigma_i^n - \bar{\sigma}) HH'(x - c_2 t) + \bar{u}(\sigma_i^n - \bar{\sigma}) H'(x - c_2 t)$$

and $\int HH'(\xi) \, d\xi = \left[\frac{H^2}{2}\right]_{-\infty}^{+\infty} = \frac{1}{2}.$

Problem 3.5.4. Consider the system

$$\begin{cases} u_t + uu_x \approx \sigma_x \\ \sigma_t + u\sigma_x \approx 0 \end{cases}$$

and travelling wave solutions of the form

$$\begin{cases} u(x,t) = \Delta u \ H(x - ct) + u_\ell \\ \sigma(x,t) = \Delta\sigma \ K(x - ct) + \sigma_\ell \end{cases}$$

with $H = f(K)$, H and K Heaviside generalized functions, where f is a $\mathscr{C}^\infty$ function of one real variable such that $f(0) = 0$ and $f(1) = 1$.

Question 1. Write the jump conditions in function of f

Question 2. Prove that the solution of the Riemann problem (by shocks),

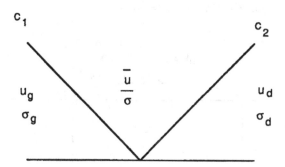

i.e. the calculation of $\bar{u}$, $\bar{\sigma}$, c_1, c_2 as a function of u_g, σ_g, u_d, σ_d, can be reduced to the solution of the system

$$\begin{cases} X^2 - (u_g + u_d) X + \dfrac{(u_g)^2 + (u_d)^2}{2} + \dfrac{\sigma_g - \sigma_d}{2a} = 0 \\ 2aX > \dfrac{1}{2} (u_g - u_d) + a (u_g + u_d) \end{cases}$$

where $X = \bar{u}$ and $a = \dfrac{1}{2} - \displaystyle\int_0^1 f(\mu) \, d\mu$, if $a \neq 0$. What in case $a = 0$?

Question 3. One considers Godunov's scheme, see 3.5.3 in the following case $u^n_{i-1} = 2$, $\sigma^n_{i-1} = 0$, $u^n_i = 1$, $\sigma^n_i = 1$, $u^n_{i+1} = 0$, $\sigma^n_{i+1} = 2$, and $a = \frac{1}{5}$; Compute u^{n+1}_i, σ^{n+1}_i as a function of $r = \frac{\Delta t}{\Delta x}$ (for $r > 0$ small enough : stability condition).

Question 4. Compute u^{n+1}_i, σ^{n+1}_i in the same conditions by integrating the equations in the cell.

Answers. In question 3 one finds $u^{n+1}_i = 1 + 2,5\, r$, $\sigma^{n+1}_i = 1 - 2,38\, r$; same results in question 4.

Problem 3.5.5 (Lax-Friedrich's scheme for nonconservative systems)
Consider the system

$$\begin{cases} u_t + uu_x \approx \sigma_x \\ \sigma_t + u\sigma_x \approx u_x \end{cases}$$

and the following configuration of meshes ($\Delta t = rh$)

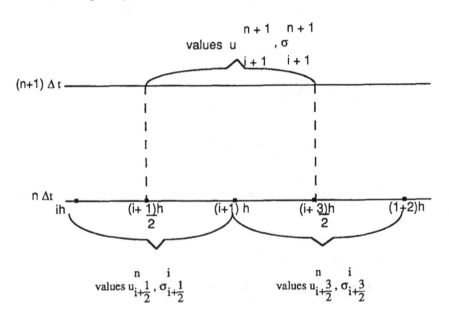

Question 1. Integrating the system in the cell $(i + \frac{1}{2}) h \le x \le (i + \frac{3}{2}) h$, $n \, \Delta t \le t \le (n+1) \, \Delta t$ compute u_{i+1}^{n+1} as a function of $u_{i+\frac{1}{2}}^n$, $\sigma_{i+\frac{1}{2}}^n$, $u_{i+\frac{3}{2}}^n$, $\sigma_{i+\frac{3}{2}}^n$, r.

Question 2. Assume the Riemann problem at $x = (i + 1) h$, $t = n \, \Delta t$ is solved by two shock waves and assume u, σ are represented by the same Heaviside function on a shock (cf. Problem 3.5.3). Assume the shocks do not meet the vertical hedges of the cell (stability condition). Compute σ_{i+1}^{n+1} as a function of $u_{i+\frac{1}{2}}^n$, $\sigma_{i+\frac{1}{2}}^n$, $u_{i+\frac{3}{2}}^n$, $\sigma_{i+\frac{3}{2}}^n$, r.

Numerical Question 3. Program the Lax-Friedrichs type scheme so obtained and compare with the exact solution.

Sketch of answers

1) $u_{i+1}^{n+1} = \frac{1}{2} (u_{i+\frac{1}{2}}^n + u_{i+\frac{3}{2}}^n) - r [\frac{1}{2} (u_{i+\frac{3}{2}}^n)^2 - \frac{1}{2} (u_{i+\frac{1}{2}}^n)^2 - \sigma_{i+\frac{3}{2}}^n + \sigma_{i+\frac{1}{2}}^n]$

2) $\sigma_{i+1}^{n+1} = \frac{1}{2} (\sigma_{i+\frac{1}{2}}^n + \sigma_{i+\frac{3}{2}}^n) - r (u_{i+\frac{1}{2}}^n - u_{i+\frac{3}{2}}^n) - \frac{r}{2} (\bar{\sigma} - \sigma_{i+\frac{1}{2}}^n) (\bar{u} + u_{i+\frac{1}{2}}^n) +$

$$+ \frac{r}{2} (\bar{\sigma} - \sigma_{i+\frac{3}{2}}^n) (\bar{u} + u_{i+\frac{3}{2}}^n)$$

with $\bar{u}$, $\bar{\sigma}$ given in problem 3.5.3.

3) it gives the same result.

A detailed study of nonconservative Lax-Friedrichs' schemes including the case considered in this problem can be found in Berger [1].

Problem 3.5.6 ("Delta waves solutions of conservative Riemann problems").

Question 1. Consider the system

$$\begin{cases} u_t + (\frac{1}{2} u^2)_x \approx 0 \\ v_t + (uv)_x \approx 0 \end{cases}$$

with the initial conditions $u_0(x) = u_\ell$ and $v_0(x) = v_\ell$ if $x < 0$, $u_0(x) = u_r$ and $v_0(x) = v_r$ if $x > 0$ (u_ℓ, v_ℓ, u_r, $v_r \in \mathbb{R}$ are arbitrary). Under what condition on these four numbers is it possible to find a solution in form of a shock wave (propagating with constant speed $c = \frac{1}{2}(u_r + u_\ell)$ from the first equation) ?

Question 2. We consider the case $u_\ell = 1$, $u_r = -1$. Find a solution v of the form

$$v(x,t) = \alpha + \beta Y(x - ct) + \gamma t\, \delta(x - ct)$$

where Y = an Heaviside function and δ = a Dirac function. Compute the numbers c, α, β, γ.

Question 3. We consider the general case u_ℓ, u_r arbitrary. Setting
$$u(x,t) = u_\ell + \Delta u\, H(x - ct)$$
find a solution of the form in question 2 for v ; What is the needed connection between H and δ ?

Question 4. Replace the second equation by
$$v_t + (u^2 v)_x \approx 0.$$
For simplicity consider first the case $u_\ell = v_\ell = 0$. Find solutions as in question 3 and 4 ; what is the needed connection between H and δ ? One may also consider u_ℓ, v_ℓ arbitrary, and also an equation
$$v_t + (\varphi(u) \cdot v)_x \approx 0.$$

Sketch of answers.
1) $(u_r - u_\ell)(v_r + v_\ell) = 0$
2) $H\delta \approx \frac{1}{2}\delta$
3) other formulas relating H and δ.

Problem 3.5.7 "Delta wave solutions of conservative Riemann problems"
The following system has been considered in Baraille-Bourdin-Dubois-Le Roux [1]

$$(E) \qquad \begin{cases} \rho_t + (\rho u)_x \approx 0 \\ (\rho u)_t + (\rho u^2)_x \approx 0 \end{cases}$$

with $u(x, o) = u_g$ if $x < o$, $u(x, o) = u_d$ if $x > o$
$\rho(x, o) = \rho_g$ if $x < o$, $\rho(x, o) = \rho_d$ if $x > o$.

In the case $u_g > u_d$ numerical tests give the result (from some natural kind of viscous approximation)

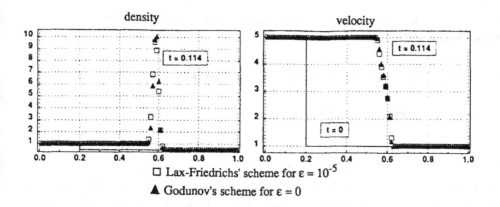

□ Lax-Friedrichs' scheme for ε = 10⁻⁵

▲ Godunov's scheme for ε = 0

One observes a peak in ρ whose height is proportional to t. This suggests to seek solutions in the form

$$(S) \quad \begin{cases} u(x,t) = u_g + \Delta u \ H(x - ct) \\ \rho(x,t) = \rho_g + \Delta\rho \ K(x - ct) + \alpha \ t \ \delta(x - ct) \end{cases}$$

H, K Heaviside generalized functions, δ a Dirac generalized function.

<u>Question</u> : compute c and α , first in the case $\rho_g = \rho_d$ then in the general case. Observe the needed relations between H and δ : is it possible to take $\delta = H'$ in the particular case $\rho_g = \rho_d$? , in the general case $\rho_g \neq \rho_d$?

<u>Answer</u> : $c = \dfrac{\sqrt{\rho_g}\,u_g + \sqrt{\rho_d}\,u_d}{\sqrt{\rho_g} + \sqrt{\rho_d}}$, $\alpha = -\sqrt{\rho_g\rho_d}\,\Delta u$; $H\delta \approx \frac{1}{2}\delta$ and $H^2\delta \approx \frac{1}{4}\delta$ if $\rho_g = \rho_d$, different values if $\rho_g \neq \rho_d$, $\delta = H'$ impossible in the general case.

Chapter 4. Jump formulas for systems in nonconservative form. New numerical methods.

§4.1 A MIXED WEAK-STRONG FORMULATION ENSURING UNIQUENESS

In chapter 3 it has been proved that the system

(1)
$$\begin{cases} u_t + uu_x \approx \sigma_x \\ \sigma_t + u\sigma_x \approx k^2 u_x \end{cases}$$

has an infinite number of possible jump conditions depending on an arbitrary parameter A (11a) b) c) in chap 3). It has also been mentioned that if each equation in (1) is stated with the (strong) equality in $\mathcal{G}$ then (1) has no solution in the requested form of a travelling wave. Both statements are not convenient to obtain existence and uniqueness at the same time. Here we consider the two intermediate statements in which one equation is stated with the association and the other one with the equality ; these statements are :

(1')
$$\begin{cases} u_t + uu_x = \sigma_x \\ \sigma_t + u\sigma_x \approx k^2 u_x \end{cases}$$
(1")
$$\begin{cases} u_t + uu_x \approx \sigma_x \\ \sigma_t + u\sigma_x = k^2 u_x. \end{cases}$$

4.1.1. Jump conditions for system (1'). At first notice that system (1') is equivalent to a system in conservation form. Indeed the first equation gives
$$u\sigma_x = uu_t + u^2 u_x.$$
Then the second equation becomes
$$\sigma_t + uu_t + u^2 u_x \approx k^2 u_x.$$
Setting $q = \sigma + \dfrac{u^2}{2}$, (1') is equivalent to

(2)
$$\begin{cases} u_t + (u^2 - q)_x = 0 \\ q_t + [\dfrac{u^3}{3} - k^2 u]_x \approx 0. \end{cases}$$

Then the jump conditions of (2) are nonambiguous (classical Rankine Hugoniot jump conditions, or merely put u, σ of the form (3) into (2)). This method is left to the reader since we prefer to use a method which works also in the nonconservative case (and gives the same result). Let us seek travelling waves of the form

(3)
$$\begin{cases} u(x,t) = \Delta u\, H(x - ct) + u_1 \\ \sigma(x,t) = \Delta\sigma\, K(x - ct) + \sigma_1 \end{cases}$$

with H, K Heaviside generalized functions. To complement the study in 3.3.6 we have to use the (strong) equality in the first equation of (1'). This equation and (3) give
$$- c\Delta u H' + (\Delta u H + u_1)\Delta u H' = \Delta\sigma K'.$$

Using (chap 3 (11a)) one obtains

$$K' = [1 - \frac{(\Delta u)^2}{2\Delta\sigma}] H' + \frac{(\Delta u)^2}{\Delta\sigma} HH'.$$

Thus

$$HK' = [1 - \frac{(\Delta u)^2}{2\Delta\sigma}] HH' + \frac{(\Delta u)^2}{\Delta\sigma} H^2 H'.$$

Using (chap 3 (3) and (11b)) one obtains

$$A = \frac{1}{2}[1 - \frac{(\Delta u)^2}{2\Delta\sigma}] + \frac{1}{3}\frac{(\Delta u)^2}{\Delta\sigma}$$

i.e.

(4) $$A = \frac{1}{2} + \frac{1}{12}\frac{(\Delta u)^2}{\Delta\sigma}$$

which, in the case of system (1'), resolves the ambiguity in chap 3 11a) b) c).

4.1.2. Jump conditions for system (1"). In the present case, we have to use the strong equality in the second equation. From (3) this equation gives the relation

$$- c\Delta\sigma K' + (\Delta uH + u_l)\Delta\sigma K' = k^2 \Delta uH'.$$

From (chap 3 (11a)) this gives the relation

$$[-\frac{1}{2} + \frac{\Delta\sigma}{(\Delta u)^2} + H]K' = \frac{k^2}{\Delta\sigma} H'.$$

If $-\frac{1}{2} + \frac{\Delta\sigma}{(\Delta u)^2} > 0$ or $-\frac{1}{2} + \frac{\Delta\sigma}{(\Delta u)^2} < 1$ and assuming $0 \le H \le 1$ then this relation can be rewritten as

(6) $$K' = \frac{k^2}{\Delta\sigma} \frac{H'}{H - \frac{1}{2} + \frac{\Delta\sigma}{(\Delta u)^2}}.$$

Multiplying (6) by H (at this point we use the fact that (6) is stated with equality in $\mathcal{G}$) and using $HK' \approx A\delta$ we get

$$A = \frac{k^2}{\Delta\sigma} \int_{-\infty}^{+\infty} \frac{H(\xi)H'(\xi)}{H(\xi) - \frac{1}{2} + \frac{\Delta\sigma}{(\Delta u)^2}} d\xi = \frac{k^2}{\Delta\sigma} \int_0^1 \frac{udu}{u - \frac{1}{2} + \frac{\Delta\sigma}{(\Delta u)^2}}$$

i.e.

(7) $$A = \frac{k^2}{\Delta\sigma} \{1 + (-\frac{1}{2} + \frac{\Delta\sigma}{(\Delta u)^2}) Log (\frac{-\frac{1}{2} + \frac{\Delta\sigma}{(\Delta u)^2}}{\frac{1}{2} + \frac{\Delta\sigma}{(\Delta u)^2}})\}.$$

We can directly obtain the jump condition from an integration of (6) :

$$K = \frac{k^2}{\Delta\sigma} Log |1 + \frac{H}{-\frac{1}{2} + \frac{\Delta\sigma}{(\Delta u)^2}}|,$$

then since K $(+\infty) = 1 = H(+\infty)$ one obtains the relation

$$1 = \frac{k^2}{\Delta\sigma} \ \text{Log} \left| 1 + \frac{1}{-\frac{1}{2} + \frac{\Delta\sigma}{(\Delta u)^2}} \right| \ ;$$

which is the jump condition corresponding to chap 3, 11b) c) with the value of A obtained in (5). Conversely one can easily prove that (1') and (1") have indeed solutions of the form (3) since this amounts to having K as a function of H. Finally :

4.1.3. Theorem. The systems (1') and (1") have nonambiguous jump conditions. These jump conditions are in general different and the values of A corresponding to these jumps are usually different from 1/2.

4.1.4. Remark. From (chap 3 (3')) the last assertion implies that H and K are different Heaviside functions when (3) represents the solutions of (1') or (1").

4.1.5 Remark. We have shown that although (1) has an infinite number of possible jump conditions, the more precise formulations (1') and (1") have nonambiguous jump conditions. Thus we have found a remedy to the ambiguity in jump conditions for systems in nonconservative form.

4.1.6. Remark. This remedy lies in some connection between the "relative strength" of the statements of the various equations of the system. In case of a system of two equations like (1), then at least intuitively the statements (1') and (1") constitute some extreme cases among an infinite family of possible relative strengths. A detailed study of this intuitive idea would require consideration of different levels of association ; this has not been done up to now.

§4.2. DOUBLE SCALE NUMERICAL SCHEMES.

When one has the jump conditions by algebraic formulas one can easily compute the solution of the Riemann problem. More important is the solution of the Cauchy problem. This has to be done numerically. In this section we present a method to construct numerical schemes for the solution of (1') and (1"). The essence of this method is very clear since it relies exactly on the concepts of $=$ and $\approx$ in $\mathcal{G}$. Usually the width of the shocks involve 3 to 4 meshes. This is not sufficient to register the equation on the jump and so this corresponds to the idea of the concept of association (i.e. equations exactly valid on both sides of the shock, and some imprecision on what happens inside the width of the shock). The equation stated with (strong) equality in $\mathcal{G}$ should be precisely written down inside the width of the shock wave. The natural idea is to use a smaller scale (for instance divide any of the above meshes into 4 smaller meshes) specifically for this equation. This gives a double scale method : a "large scale" for the equation stated with association and a "small scale" for the equation stated with equality in $\mathcal{G}$. In this section we expose this double scale method for (1') and (1"), and we check that it gives back (numerically) the jump conditions obtained in 4.1.1 and 4.1.2. We start from the one scale method (S_0) considered in 3.4.1. As this was checked in 3.4.1 the scheme (S_0) gives the value $A = \frac{1}{2}$ for the shocks. We expose the modifications of (S_0) suggested by the above double scale method.

4.2.1. <u>Proposed numerical scheme for the solution of (1')</u>. For the space discretization the "large" meshes are divided into four equal "small" meshes ; the time discretization is only made of one scale. The index i refers to the small (space) meshes discretization. The large scale is used for the discretization of the second equation in (1') ; the values σ_i^n are defined only when i is a multiple of four.For all $i \in \mathbb{Z}$ we set

(8$_1$)
$$u_i^{n+1} = m_i^n - rm_i^n \, \frac{u_{i+1}^n - u_{i-1}^n}{2} + r(d\sigma)_i^n.$$

$(d\sigma)_i^n$ has to discretize the term σ_x but σ_i^n is defined only when i is a multiple of four ; the following natural discretization of $(d\sigma)_i^n$ has been adopted : first let us consider the case $i \geq 0$;

then if $i = 4k$, $k \in \mathbb{N}$ we set $(d\sigma)_i^n = \frac{1}{8}(\sigma_{i+4}^n - \sigma_{i-4}^n)$

if $i = 4k + 1$ we set $(d\sigma)_i^n = \frac{3}{16}(\sigma_{4k+4}^n - \sigma_{4k}^n) + \frac{1}{16}(\sigma_{4k}^n - \sigma_{4k-4}^n)$

if $i = 4k + 2$ we set $(d\sigma)_i^n = \frac{1}{4}(\sigma_{4k+4}^n - \sigma_{4k}^n)$

if $i = 4k + 3$ we set $(d\sigma)_i^n = \frac{3}{16}(\sigma_{4k+4}^n - \sigma_{4k}^n) + \frac{1}{16}(\sigma_{4k+8}^n - \sigma_{4k+4}^n)$.

Symmetric definitions are adopted for the case $i \leq 0$. For those i which are integer multiples of four, (8$_1$) is complemented by

(8$_2$)
$$\sigma_i^{n+1} = \frac{\sigma_{i+4}^n + 2\sigma_i^n + \sigma_{i-4}^n}{2} - rm_i^n \frac{\sigma_{i+4}^n - \sigma_{i-4}^n}{8} + rk^2 \frac{u_{i+1}^n - u_{i-1}^n}{2}.$$

Various slightly different discretizations have been attempted, in particular for the term $(d\sigma)_i^n$ in (8$_1$) :

all of them work but give more or less precise results. The key of the method lies clearly in the idea of double scaling - as suggested by our mathematical theory - and not in details of programation.

4.2.2. <u>Proposed numerical scheme for the solution of (1")</u>. In this case the values of u_i^n are defined only when i is a multiple of four (large scale). Thus the scheme becomes : for $i \in \mathbb{Z}$

$$(9_1) \qquad \sigma_i^{n+1} = g_i^n - r(Mu)_i^n \frac{\sigma_{i+1}^n - \sigma_{i-1}^n}{2} + rk^2(du)_i^n$$

where $(Mu)_i^n$ and $(du)_i^n$ are defined similarly as before, due to the fact that u_i^n is defined only for i

multiple of four : for $i \geq 0$

if $i = 4k$, set

$$(du)_i^n = \frac{1}{8}(u_{i+4}^n - u_{i-4}^n) \text{ and } (Mu)_i^n = \frac{1}{4}(u_{i+4}^n + 2u_i^n + u_{i-4}^n)$$

if $i = 4k + 1$, set

$$(du)_i^n = \frac{3}{16}(u_{4k}^n - u_{4k-4}^n) \text{ and } + \frac{1}{16}(u_{4k+4}^n - u_{4k}^n), (Mu)_i^n = \frac{1}{4}(u_{4k+4}^n + 2u_{4k}^n + u_{4k-4}^n),$$

if $i = 4k + 2$, set

$$(du)_i^n = \frac{1}{4}(u_{4k+4}^n + u_{4k}^n) \text{ and } (Mu)_i^n = \frac{1}{2}(u_{4k+4}^n + u_{4k}^n)$$

if $i = 4k + 3$, set

$$(du)_i^n = \frac{3}{16}(u_{4k+4}^n - u_{4k}^n) + \frac{1}{16}(u_{4k+8}^n - u_{4k+4}^n), (Mu)_i^n = \frac{1}{4}(u_{4k+8}^n + 2u_{4k+4}^n + u_{4k}^n).$$

For i multiple of 4, the first equation in (1") is discretized by

$$(9_2) \quad u_i^{n+1} = \frac{u_{i+4}^n + 2u_i^n + u_{i-4}^n}{4} - r\frac{u_{i+4}^n + 2u_i^n + u_{i-4}^n}{4} \cdot \frac{u_{i+4}^n + u_{i-4}^n}{8} + r\frac{\sigma_{i+4}^n - \sigma_{i-4}^n}{8}.$$

Numerical results are shown in figures 1 and 2 below.

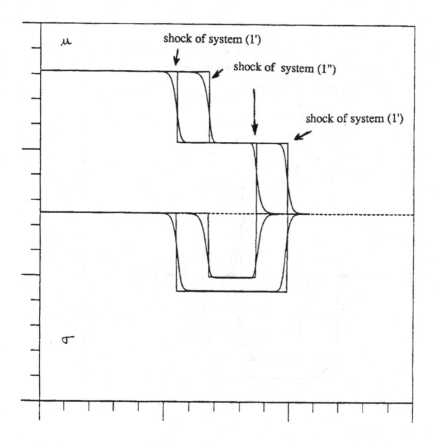

Figure 1. Theoretical and numerical results for systems (1') and (1") with initial values $u_l = 3$, $\sigma_l = 0 = u_r = \sigma_r$ and for $k^2 = 1$. The numerical results follow from the schemes $8_{1,2}$ and $9_{1,2}$. The theoretical results for (1') are quite different from those for (1"). The numerical results coincide with the theoretical ones.

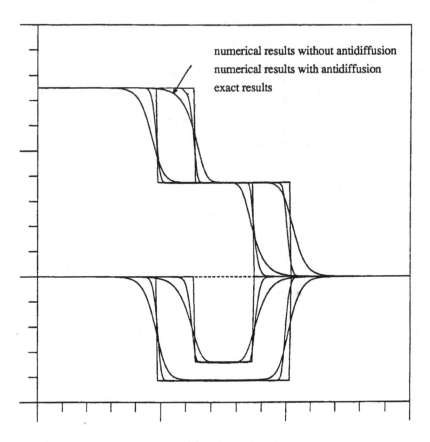

numerical results without antidiffusion
numerical results with antidiffusion
exact results

Figure 2. In addition to the curves in fig. 1 curves are shown obtained from the antidiffusion technique of 3.4.1 adapted to the schemes $8_{1,2}$ and $9_{1,2}$. One observes that antidiffusion techniques can be adapted to the double scale method and that they produce efficient results.

§4.3. A GENERAL METHOD FOR THE REMOVAL OF THE AMBIGUITY.

It has been found in §3.3, §3.4 that systems in nonconservative form usually have an infinite number of possible jump conditions, depending on one (or several) arbitrary real parameters. A way to ensure uniqueness of jump condition has been found in §4.1 : could we use it to resolve the ambiguity in systems of equations from physics ? Of course such a resolution has to be decided on physical ground. Indeed there are several ways to select one solution (the solutions of (1') and (1") are different). The idea to state some equations with the equality in $\mathcal{G}$ (strong formulation of these equations) and other equations with the association (weak formulation) can be easily justified from very simple physical arguments. The method we propose is based on the difference between the basic laws of classical physics (balance of mass, of momentum, of energy and of electric charge) and the constitutive equations. Since the shock waves have physically some nonzero width (of the order of magnitude of a few mean free paths, at least) we postulate that the basic laws of physics are still valid within this width. On the other hand, the constitutive equations are no more than results of measurements in a given circumstance ; these measurements have never been done in a state of extremely fast deformation as the one inside the width of the shock waves. Therefore they are valid on both sides of the shock, and (possibly) not within its width. This difference between these two concepts (basic laws and constitutive equations) corresponds rather clearly to the difference between the (strong) equality in $\mathcal{G}$ and the association (remember our comparison of H and H^n in §3.1). Thus our method will naturally consist in stating the basic laws of physics with the (strong) equality in $\mathcal{G}$ and the constitutive equations with the association. In the case only one constitutive equation is involved in the system we shall ascertain on examples that this method solves the ambiguity. However in case the system contains more than one constitutive equation there usually subsists an ambiguity that should be solved from supplementary physical arguments. In the sequel of this chapter we shall consider only the "nice" case of only one constitutive equation. Any physical method has its own limitations : in the equations which express basic laws of physics various effects such as viscosity, heat conduction, external forces, ... have been neglected ; in some cases this might invalidate their statement with (strong) equality in $\mathcal{G}$.

§4.4. STUDY OF A SYSTEM OF THREE EQUATIONS.

The following is a simplified model of the system of elasticity (see chap 2). Hooke's law is expressed in terms of the stress σ in the form $\frac{d}{dt}\sigma = k^2 u_x$ where $\frac{d}{dt} = \frac{\partial}{\partial t} + u\frac{\partial}{\partial x}$ is the material derivative

(10)
$$\begin{cases} \rho_t + (\rho u)_x \approx 0 & \text{balance of mass} \\ (\rho u)_t + (\rho u^2)_x \approx \sigma_x & \text{balance of momentum} \\ \sigma_t + u\sigma_x \approx k^2 u_x & \text{Hooke's law} \end{cases}$$

ρ = density, u = velocity, k^2 = constant. System (10) is stated with three associations since we know this statement is a faithful generalization of the concept of weak solutions of systems in conservative form.

4.4.1. <u>Jump conditions of system (10)</u>. We seek travelling waves solutions of (10) of the form

(11) $$\begin{cases} u(x,t) = \Delta u H(x - ct) + u_1 \\ \sigma(x,t) = \Delta\sigma K(x - ct) + \sigma_1 \\ \rho(x,t) = \Delta\rho L(x - ct) + \rho_1 \end{cases}$$

with H, K and L three Heaviside generalized functions. Putting (11) into the first equation of (10) we get (assuming $\Delta\rho \neq 0$)

(12a) $$c - u_1 = \Delta u + \rho_1 \frac{\Delta u}{\Delta\rho}.$$

The second equation of (10) gives

(12b) $$(c - u_1 - \Delta u)(u_1\Delta\rho + \rho_1\Delta u + \Delta\rho\Delta u) = u_1\rho_1\Delta u - \Delta\sigma.$$

Note that (12a) b)) are exactly the classical Rankine Hugoniot jump conditions, since these equations are in conservative form. As in 3.3.6 the last equation in (10) gives

(12c) $$c - u_1 = A\Delta u - k^2\frac{\Delta u}{\Delta\sigma}$$

and

(12d) $$HK' \approx A\delta.$$

12a) b) c) can be rewritten as :

(13) $$\begin{cases} c = u_1 + A\Delta u - k^2\frac{\Delta u}{\Delta\sigma} \\ k^2\frac{\Delta u}{\Delta\sigma} - \frac{1}{2}\left[\frac{1}{\rho_1} + \frac{1}{\rho_r}\right]\frac{\Delta u}{\Delta\sigma} = \left[A - \frac{1}{2}\right]\Delta u \\ \rho_1\rho_r(\Delta u)^2 = -\Delta\sigma\Delta\rho. \end{cases}$$

As usual we find that the jump conditions of (10) depend on an arbitrary parameter : the real number A defined by (12d).

4.4.2. <u>Resolution of the ambiguity</u>. The idea exposed in §4.3 and the study in § 4.1 suggest to state system (10) in the more precise form

(14) $$\begin{cases} \rho_t + (\rho u)_x = 0 \\ (\rho u)_t + (\rho u^2)_x = \sigma_x \\ \sigma_t + u\sigma_x \approx k^2 u_x. \end{cases}$$

The first two equations in (14) are equivalent to

$$(15) \quad \begin{cases} \rho_t + (\rho u)_x = 0 \\ u_t + u u_x = \dfrac{1}{\rho} \sigma_x \end{cases}$$

(since $\rho \neq 0$). It is convenient to set $v = \dfrac{1}{\rho}$ (v is called the specific volume). Then (15) takes the form

$$(15') \quad \begin{cases} v_t + u v_x - v u_x = 0 \\ u_t + u u_x - v \sigma_x = 0. \end{cases}$$

Let us restate (11) in the form

$$(11') \quad \begin{cases} u(x,t) = \Delta u H(x - ct) + u_l \\ \sigma(x,t) = \Delta\sigma K(x - ct) + \sigma_l \\ v(x,t) = \Delta v M(x - ct) + v_l \end{cases}$$

with H, K, M $\approx$ Y. The first equation in (15') gives

$$(16) \quad (c - u_l - \Delta u H)M' + \Delta u H'M + \Delta u \frac{v_l}{\Delta v} H' = 0.$$

The jump condition of the first equation in (14) is

$$\frac{c - u_l}{\Delta u} = \frac{\rho_r}{\Delta\rho} \quad \text{with } \rho_r = \Delta\rho + \rho_l$$

i.e.

$$(14') \quad \frac{c - u_l}{\Delta u} = -\frac{v_l}{\Delta v}.$$

Then (16) gives

$$(16') \quad \left[\frac{v_l}{\Delta v} + H\right] M' = \left[\frac{v_l}{\Delta v} + M\right] H'.$$

Putting (11') into the second equation of (15') one obtains :

$$(17) \quad \frac{c - u_l}{\Delta u} H' - H H' + (\Delta v M + v_l) \frac{\Delta\sigma}{\Delta u^2} K' = 0.$$

The jump condition of the second equation in (14) gives

$$(14'') \quad \Delta v = \frac{(\Delta u)^2}{\Delta\sigma}.$$

(17), (14') and (14") give

$$(17') \quad \left[\frac{v_l}{\Delta v} + H\right] H' = \left[\frac{v_l}{\Delta v} + M\right] K'.$$

Setting $\alpha = \dfrac{v_l}{\Delta v} > 0$ then (16') (17') are the system

$$(18) \quad \begin{cases} (\alpha + H)M' = (\alpha + M)H' \\ (\alpha + H)H' = (\alpha + M)K'. \end{cases}$$

(18) can be rewritten as

$$\begin{cases} (\alpha + H)M' - H'H - \alpha H' = 0 \\ \quad K' = \dfrac{\alpha + H}{\alpha + M} H'. \end{cases}$$
(18')

Since H, M are null on $(-\infty, 0[$, identical to 1 on $]0,+\infty)$, an application of the classical formula for the solution of the ordinary differential equation

$$a(x)y' + b(x)y + c(x) = 0$$

allows to compute M as a function of H from the first equation of (18'). One finds M = H. This method relies upon the extension to $\mathcal{G}$ of the classical study of ordinary differential equations of the above kind. One can check that in this case the classical formula makes sense and provides a unique solution in the sense of equality in $\mathcal{G}$. This study can be considered as a particular case of a much deeper study of linear hyperbolic systems with coefficients in $\mathcal{G}$ (Oberguggenberger [6]). There one proves the uniqueness of solutions of the Cauchy problem ; this argument of uniqueness gives at once (without any computation) the result that M = H. Then the second equation in (18') gives K = H. We have resolved the ambiguity :

$$HK' = HH' \approx \frac{1}{2}\delta \text{ [and thus A } = \frac{1}{2}\text{] and obtained :}$$

4.4.3. Theorem - The system of two equations and three unknowns

(19)
$$\begin{cases} \rho_t + (\rho u)_x = 0 \\ (\rho u)_t + (\rho u^2)_x - \sigma_x = 0 \end{cases}$$

is equivalent to the system $(v = \dfrac{1}{\rho})$

(20)
$$\begin{cases} v_t + uv_x - vu_x = 0 \\ u_t + uu - v\sigma_x = 0. \end{cases}$$

Further, travelling waves of the form

(21)
$$w(x,t) = \Delta w H_w(x - ct) + w_\ell,$$

$w = v, u, \sigma$ successively $(\Delta w, c, w_\ell \in \mathbb{R}, H_w$ a Heaviside generalized function), are solutions of (20) if and only if $H_v = H_u = H_\sigma$, plus the classical jump conditions of (19).

The necessary part of the proof is the one given above. The converse is an easy verification.

4.4.4. Numerical schemes for the solution of (14). One can, of course, use the double scale method of §4.2. But also we have obtained the remarkable result that when the system is stated in the nonconservative form

(22)
$$\begin{cases} v_t + uv_x - vu_x = 0 \\ u_t + uu_x - v\sigma_x = 0 \\ \sigma_t + u\sigma_x \approx k^2 u_x \end{cases}$$

then the Heaviside functions of u, v, σ are the same (this follows from the fact the first two equations are stated with the equality in $\mathcal{G}$). In this case one often observes that adaptations of classical one-scale schemes used in the conservative case give the correct result. This observation, which has not been fully explained, comes probably from the fact that, for reasons of simplicity, all variables are usually treated in the same way in these schemes. Numerical examples will be given in the sequel of this chapter, in the basically important case of fluid dynamics.

4.4.5. Simplification of (14) in case ρ is approximately constant.

The second equation in (14) gives

$$\rho_t u + \rho u_t + (\rho u)_x u + (\rho u)u_x = \sigma_x.$$

From the first equation in (14) - which may be multiplied by u since it is stated with equality in $\mathcal{G}$ - one obtains

$$u_t + u u_x = \frac{\sigma_x}{\rho}.$$

Assuming that ρ is approximately constant (equal to some value $\rho_0 \in \mathbb{R}$) then (14) becomes the system

(23)
$$\begin{cases} u_t + u u_x = \dfrac{\sigma_x}{\rho_0} \\ \sigma_t + u \sigma_x \approx k^2 u_x \end{cases}$$

previously studied in §4.1. However the replacement of ρ by ρ_0 appears to be abusive : since $|\sigma_x|$ is very large (infinite) on the shock, the replacement of $\dfrac{\sigma_x}{\rho}$ by $\dfrac{\sigma_x}{\rho_0}$ might modify the equation significantly. Thus it would be more reasonable to state the simplified system in the weaker form

(23')
$$\begin{cases} u_t + u u_x \approx \dfrac{\sigma_x}{\rho_0} \\ \sigma_t + u \sigma_x \approx k^2 u_x . \end{cases}$$

But from §3.4 one knows that the jump conditions for (23') are ambiguous. Setting $\rho_0 = 1$ for simplicity these jump conditions are (11 a, b, c) in chap. 3. They should be an approximation, for ρ remaining close to 1, of those of (14) : i.e. (13) with $A = \frac{1}{2}$. This is a way to resolve the ambiguity in (23') different from those proposed in §4.1 ; it has been adopted in Le Roux-De Luca [1] ; Numerical schemes corresponding to these values (besides the very simple one in §3.4) are given in Cauret [1], Cauret-Colombeau-Le Roux [1], Le Roux-De Luca [1].

§4.5. A NONCONSERVATIVE FORMULATION OF FLUID DYNAMICS. We use the classical notations ρ = density, u = velocity, p = pressure, e = density of total energy, I = density of internal energy,

$$(24) \qquad\qquad e = I + \frac{u^2}{2}$$

$v = \dfrac{1}{\rho}$ = specific volume, ϕ = a C^{∞} function of the variables ρ and I. We assume that the equation of state $p = \phi(\rho, I)$ can be locally stated as $I = \psi(\rho, p)$ with ψ a C^{∞} function (this holds in practice). Then the system of fluid dynamics can be stated as

$$\begin{cases} \rho_t + (\rho u)_x \approx 0 & \text{balance of mass} \\ (\rho u)_t + (\rho u^2)_x + p_x \approx 0 & \text{balance of momentum} \\ (\rho e)_t + (\rho e u + p u)_x \approx 0 & \text{balance of energy} \\ p \approx \phi(\rho, I) \text{ or } I \approx \psi(\rho, p) & \text{equation of state.} \end{cases}$$

The method exposed at the beginning of this chapter consists in stating the first three equations in (25) with the equality in $\mathcal{G}$. Indeed it will suffice for the sequel to state only the first two equations with equality in $\mathcal{G}$. Thus the system we shall consider is the following

$$(25') \qquad \begin{cases} \rho_t + (\rho u)_x = 0 \\ (\rho u)_t + (\rho u^2)_x + p_x = 0 \\ (\rho e)_t + (\rho e u + p u)_x \approx 0 \\ I \approx \psi(\rho, I) \text{ and } e = I + \frac{u^2}{2}. \end{cases}$$

From theorem 4.4.3 the first two equations are equivalent to

$$(26) \qquad \begin{cases} v_t + u v_x - v u_x = 0 \\ u_t + u u_x + v p_x = 0 \end{cases}$$

(replace σ by $- p$). Now we eliminate I in the last line of (25') :

$$(27) \qquad\qquad e \approx \psi(\rho, p) + \frac{u^2}{2} .$$

We want to replace e by this value in the equation expressing the balance of energy ; thus we need an expression for ρe and $\rho e u$. The association is incompatible with the multiplication so that this looks a priori impossible. We are interested in the case in which ρ, e, p u $\in \mathcal{G}(\mathbb{R}^2)$ have the macroscopic aspect of piecewise continuous functions (§3.2 : i.e. this means that they are associated in $(\mathcal{G}(\mathbb{R}^2))$ to piecewise continuous functions $\bar{\rho}, \bar{e}, \bar{p}, \bar{u}$ respectively). Thus (27) can be interpreted as

(27')
$$\bar{e} = \psi(\bar{\rho}, \bar{p}) - \frac{\bar{u}^2}{2}$$

in the classical sense of the equality of classical functions (i.e. this equality holds for all $(x,t) \in \mathbb{R}^2$ except on a set of zero Lebesgue measure : the set of discontinuities). (27') implies, in this classical sense,

(27")
$$\bar{\rho}\,\bar{e} = \bar{\rho}\psi(\bar{\rho}, \bar{p}) + \bar{\rho}\,\frac{\bar{u}^2}{2} \;; \quad \bar{\rho}\,\bar{e}\,\bar{u} = \bar{\rho}\,\bar{u}\,\psi(\bar{\rho}, \bar{p}) + \bar{\rho}\,\frac{\bar{u}^3}{2}.$$

We assume that ρ, e, p, u have respective representatives (ρ_ε), (e_ε), (p_ε), (u_ε) that are bounded above in absolute value (on any bounded subset of $\mathbb{R}^2$) independently of ε (as in c) of (3.3.1)).

$\rho_\varepsilon(x,t) \to \bar{\rho}(x,t)$ when $\varepsilon \to 0$ for almost all (x,t), and same for the other physical variables ; further one can apply the dominated convergence theorem : for any $\varphi \in \mathcal{D}(\mathbb{R}^2)$

$$\int (\rho_\varepsilon e_\varepsilon - \rho_\varepsilon\psi(\rho_\varepsilon,p_\varepsilon) - p_\varepsilon\frac{u_\varepsilon^2}{2})\,(x,t)\,\varphi\,(x,t)dxdt \to 0 \text{ as } \varepsilon \to 0.$$

One obtains

(27''')
$$\rho e \approx \rho\psi(\rho,p) + \rho\frac{u^2}{2} \;, \rho e u \approx \rho u \psi(\rho,p) + \rho\frac{u^3}{2}.$$

<u>Remark.</u> One can prove easily that the new product in $\mathcal{G}$ of piecewise continuous functions is associated to the classical product (see chap 8); the passage from (27) and (27''') is an aspect of this basic fact ; it would be false in general in case ρ, e, p would be more singular.

(27''') and the third equation in (25') yield :

$$[\rho_t\psi + \rho\rho_t D_1\psi + \rho p_t D_2\psi + \rho_t\frac{u^2}{2} + \rho u u_t] +$$

$$+ [\rho_x u\psi + \rho u_x\psi + \rho\rho_x u D_1\psi + \rho u p_x D_2\psi + \rho_x\frac{u^3}{2} + \frac{3}{2}\rho u^2 u_x] + [p_x u + p u_x] \approx 0.$$

Simplifications and use of the first two equations of (25') give, after some calculations :

(28)
$$- \rho^2 u_x D_1\psi + \rho(p_t + u p_x)\,D_2\psi + p u_x \approx 0.$$

Defining a C^∞ function φ by

(29)
$$\varphi(v,p) = \psi[\frac{1}{v}, p] = \psi(\rho,p)$$

one has

$$\frac{\partial\varphi}{\partial v}(v,p) = -\frac{1}{v^2}D_1\psi(\rho,p) \text{ and } \frac{\partial\varphi}{\partial p}(v,p) = D_2\psi(\rho,p).$$

Then (28) becomes

(28')
$$\frac{1}{v}\cdot\frac{\partial\varphi}{\partial p}(v,p)\cdot p_t + \frac{u}{v}\cdot\frac{\partial\varphi}{\partial p}(v,p)\cdot p_x + [p + \frac{\partial\varphi}{\partial v}(v,p)]u_x \approx 0.$$

We have obtained :

4.5.1 <u>Theorem. The piecewise C^∞ solutions of the system (25') of fluid dynamics are solutions of the system</u>

$$(30) \quad \begin{cases} v_t + uv_x - vu_x = 0 \\ u_t + uu_x + vp_x = 0 \\ [\frac{1}{v} \frac{\partial \varphi}{\partial p}(v,p)]p_t + [\frac{u}{v} \frac{\partial \varphi}{\partial p}(v,p)]p_x + [p + \frac{\partial \varphi}{\partial v}(v,p)]u_x \approx 0. \end{cases}$$

Further from theorem 4.4.3 we recall the remarkable result that in the shocks the variables v, u, p are represented by the same Heaviside generalized function.

4.5.2. Simplification of the third equation of (30) in particular cases. The third equation of (30) looks very complicated. Fortunately in most cases encountered in practice it takes a very simple form. In the case of aerodynamics the equation of state takes the very simple form

$$(31) \qquad p \approx (\gamma - 1) \rho I$$

$\gamma > 1$ constant. Then $\varphi(v,p) = \dfrac{vp}{\gamma - 1}$ and the third equation of (30) takes the form

$$(30') \qquad p_t + up_x + \gamma pu_x \approx 0.$$

When one uses the system of fluid dynamics to model collisions (i.e. by neglecting the elastic stage) one often uses a Mie Gruneisen equation of state

$$(32) \qquad p \approx \gamma \rho I + G(\rho)$$

with $\gamma > 0$ and G a function of ρ. (32) gives

$$\varphi(v,p) = \frac{1}{\gamma} (vp - vG(\frac{1}{v}))$$

and one obtains easily that the third equation of (30) takes the form

$$(30'') \qquad p_t + up_x + [(\gamma + 1) p + F (v)] u_x \approx 0.$$

where

$$(30''') \qquad F(v) = \frac{1}{v} G'[\frac{1}{v}] - G[\frac{1}{v}] = \rho G'(\rho) - G(\rho).$$

The systems (30) (30') and (30) (30") will be exploited numerically in the next two sections to provide new numerical schemes in aerodynamics and numerical simulation of collisions. Other Mie Gruneisen equations of state are in the form

$$(32') \qquad p \approx \alpha I + G(\rho) \qquad , \quad \alpha > 0$$

i.e.

$$I \approx \frac{1}{\alpha}(p - G(\frac{1}{v})) = \varphi(v,p)$$

since p, I, ρ have the macroscopic aspect of piecewise continuous functions, see chap. 8.

The third equation in (30) takes the form

(30⁴)

$$\frac{1}{v}p_t + \frac{u}{v}p_x + [\alpha p + \frac{1}{v^2} G'(\frac{1}{v})] u_x \approx 0.$$

Since this equation is stated with association multiplication by v is forbidden.

Remark. The transformation of the system of fluid dynamics performed in this section has been extended to the 3 dimensional case and to the 3 dimensional axisymmetric case (with the choice of variables adapted to this case) in Arnaud [1].

Research Problem. We shall show in the next sections that the nonconservative system (30, 30', 30") leads to efficient numerical methods. Thus the reader could attempt similar works on other classical conservative systems.

§ 4.6 NON CONSERVATIVE NUMERICAL SCHEMES IN FLUID DYNAMICS

In this section we consider the system of gas dynamics (30) (30') (with $\gamma = 1, 4$), i.e. the system

(33)
$$\begin{cases} v_t + uv_x - vu_x = 0 \\ u_t + uu_x + vp_x = 0 \\ p_t + up_x + \gamma pu_x \approx 0. \end{cases}$$

4.6.1. A very simple finite difference discretization of (33) : the NC1 (nonconservative 1) scheme.

The first step is the discretization step described below ; the second step is the classical antidiffusion method already used in 3.4.3. One defines the mean values

(34ₐ)
$$\begin{cases} (mu)_i^n = \frac{1}{4}(u_{i+1}^n + 2u_i^n + u_{i-1}^n) \\ (mv)_i^n = \frac{1}{4}(v_{i+1}^n + 2v_i^n + v_{i-1}^n) \\ (mp)_i^n = \frac{1}{4}(p_{i+1}^n + 2p_i^n + p_{i-1}^n). \end{cases}$$

For $i \in \mathbb{Z}$ ones computes

$$(34_b) \quad \begin{cases} v_i^{n+1} = (mv)_i^n - r\,(mu)_i^n \dfrac{v_{i+1}^n - v_{i-1}^n}{2} + r\,(mv)_i^n \dfrac{u_{i+1}^n - u_{i-1}^n}{2} \\[3ex] u_i^{n+1} = (mu)_i^n - r\,(mu)_i^n \dfrac{u_{i+1}^n - u_{i-1}^n}{2} - r\,(mv)_i^n \dfrac{p_{i+1}^n - p_{i-1}^n}{2} \\[3ex] p_i^{n+1} = (mp)_i^n - r\,(mu)_i^n \dfrac{p_{i+1}^n - p_{i-1}^n}{2} - r\gamma(mp)_i^n \dfrac{u_{i+1}^n - u_{i-1}^n}{2} \; . \end{cases}$$

One will observe in 4.6.5 below that this scheme gives the correct solution. This is closely related to the observation in § 3.4 that there the scheme (S_0) gives also the solutions corresponding to the value $A = \dfrac{1}{2}$ (fig. 7 of chap. 3)

4.6.2 The NC1 scheme with double scale.

The double scale method exposed in § 4.2 can be adapted to the NC1 scheme by using small meshes for the first two equations in (33). For $i \in \mathbb{Z}$ one computes

$$(35_a) \quad \begin{cases} v_i^{n+1} = (mv)_i^n - r\,(mu)_i^n \dfrac{v_{i+1}^n - v_{i-1}^n}{2} + r\,(mv)_i^n \dfrac{u_{i+1}^n - u_{i-1}^n}{2} \\[3ex] u_i^{n+1} = (mu)_i^n - r\,(mu)_i^n \dfrac{u_{i+1}^n - u_{i-1}^n}{2} + r\,(mv)_i^n \,(dp)_i^n. \end{cases}$$

For $i \in 4\,\mathbb{Z}$ one computes

$$(35_b) \quad p_i^{n+1} = \dfrac{p_{i+4}^n + 2p_i^n + p_{i-4}^n}{4} - r\,(mu)_i^n \dfrac{p_{i+4}^n - p_{i-4}^n}{8}$$

$$- \gamma r \dfrac{p_{i+4}^n + 2p_i^n + p_{i-4}^n}{4} \cdot \dfrac{u_{i+1}^n - u_{i-1}^n}{2} \; .$$

The quantity $(dp)_i^n$ is defined by :

$$(35_a)\begin{cases} \text{if } i = 4k \ (k \in \mathbb{N}) & dp_i^n = \frac{1}{8}(p_{i+4}^n - p_{i-4}^n) \\[2mm] \text{if } i = 4k + 1 & dp_i^n = \frac{1}{16}(p_{4k+4}^n - p_{4k}^n) + \frac{3}{16}(p_{4k}^n - p_{4k-4}^n) \\[2mm] \text{if } i = 4k + 2 & dp_i^n = \frac{1}{4}(p_{k+4}^n - p_{k-4}^n) \\[2mm] \text{if } i = 4k + 3 & dp_i^n = \frac{3}{16}(p_{4k+4}^n - p_{4k-4}^n) + \frac{1}{16}(p_{4k}^n - p_{4k-4}^n) \end{cases}$$

(symmetric definition for $i < 0$). The antidiffusion step is the same as in 3.4.3 and is applied for i multiple of 4.

It has been observed that the double scale method gives the correct result. Various splittings using the double scale method are considered in 4.7.4 .

4.6.3. A Godunov type scheme with splitting into a convection part a propagation part : the NC2 scheme.

This scheme (due to A.Y. Le Roux, see Colombeau-Le Roux [2]) is decomposed into three steps :

step 1 : treatment of the convection by the Godunov scheme.

step 2 : treatment of the propagation by a Godunov type scheme in which the flux is calculated from an approximate Riemann solver.

step 3 : antidiffusion.

4.6.3. a) Description of step 1 (convection). See Arnaud [1], Noussaïr [1], De Luca [1] ; see problem 4.7.2 for details. The convection part consists in the system

$$(36)\begin{cases} u_t + uu_x \approx 0 & (36a) \\ v_t + uv_x \approx 0 & (36b) \\ p_t + up_x \approx 0 & (36c). \end{cases}$$

One solves Burgers 'equation (36a) by the Godunov scheme and one treats (36b) (36c) in such a way that u, v, p are represented by the same Heaviside functions on a shock. On each mesh one computes the fluxes through the interfaces $x = jh$, $(j = i + \frac{1}{2})$; this gives

* if $u_{i+1}^n > 0$ and $u_i^n + u_{i+1}^n > 0$

$$u_j^n = u_i^n \ , \ v_j^n = v_i^n \ , \ p_j^n = p_i^n$$

* if $u_i^n < 0$ and $u_i^n + u_{i+1}^n < 0$

$$u_j^n = u_{i+1}^n \ , \ v_j^n = v_{i+1}^n \ , \ p_j^n = p_{i+1}^n$$

* if $u_{i+1}^n > 0$ and $u_i^n < 0$

$$u_j^n = 0$$

$$v_j^n = v_i^n + \frac{v_{i+1}^n - v_i^n}{u_{i+1}^n - u_i^n} \ (u_j^n - u_i^n) = \frac{u_{i+1}^n \ v_i^n - u_i^n \ v_{i+1}^n}{u_{i+1}^n - u_i^n}$$

$$p_j^n = p_i^n + \frac{p_{i+1}^n - p_i^n}{u_{i+1}^n - u_i^n} \ (u_j^m - u_i^m) = \frac{u_{i+1}^n \ p_i^n - u_i^n \ p_{i+1}^n}{u_{i+1}^n - u_i^n}.$$

The speeds of waves are

$$C_{L,j} = \frac{u_j^n + u_i^n}{2} \qquad \text{for the left hand side wave}$$

$$C_{R,j} = \frac{u_{i+1}^n + u_j^n}{2} \qquad \text{for the right hand side wave.}$$

Then one projects on each mesh by the usual projection operator of the L^2 norm. Setting $m = n + \frac{1}{2}$ one has

$$w_i^m = rC_{R,j-1} \ w_{j-1}^n + (1 + rC_{L,j} - rC_{R,j-1}) \ w_i^n - rC_{L,j} w_j^n$$

for $w = v, u, p$. The stability condition is

$$r \max_{i \in \mathbb{Z}} (u_i^n) \le \frac{1}{2}.$$

4.6.3. b) <u>Description of step 2 (propagation)</u>. Detailed proofs are given in 4.7.2 below. The propagation part consists in the system

(37)
$$\begin{cases} v_t - v u_x \approx 0 \\ u_t + v p_x \approx 0 \\ p_t + \gamma p u_x \approx 0 \end{cases}.$$

One computes an approximate solution of (37) on the strip $\mathbb{R} \times [t_n, t_{n+1}]$ from the values v_i^m, u_i^m, p_i^m

computed at step 1. At first one computes the fluxes through the cell interfaces. This is done through an approximate Riemann solver ; in this case the solution of the Riemann problem is made with two waves (shock or rarefaction) and a contact discontinuity. It gives a pressure p_j^m, a velocity u_j^m and

two values $v_{1,j}^m$ and $v_{2,j}^m$ of the specific volume (recall that $j = 1 + \frac{1}{2}$ and $m = n + \frac{1}{2}$). One sets

(acoustic impedance)

$$Z_{R,j}^m = \sqrt{\gamma \frac{p_{i+1}^m + p_j^m}{v_{i+1}^m + v_{2,j}^m}} \quad \text{and} \quad Z_{L,j}^m = \sqrt{\gamma \frac{p_j^m + p_i^m}{v_{1,j}^m + v_i^m}}.$$

The four numbers p_j^m, u_j^m, $v_{1,j}^m$ and $v_{2,j}^m$ are linked by the four relations (jump conditions)

(38)
$$\begin{cases} u_{i+1}^m - u_j^m = - Z_{R,j}^m (v_{i+1}^m - v_{2,j}^m) \\ p_{i+1}^m - p_j^m = - (Z_{R,j}^m)^2 (v_{i+1}^m - v_{2,j}^m) \\ u_i^m - u_j^m = Z_{L,j}^m (v_i^m - v_{1,j}^m) \\ p_i^m - p_j^m = (Z_{L,j}^m)^2 (v_i^m - v_{1,j}^m) \end{cases}.$$

The solver consists in solving this nonlinear system by Newton's iterative method (convergence is obtained in 3 or 4 iterations, with an error $\leq 10^{-4}$). The sound speeds are

$$C_{R,j}^m = Z_{R,j}^m \frac{v_{i+1}^m + v_{2,j}^m}{2} \quad , \quad C_{L,j}^m = - Z_{L,j}^m \frac{v_{1,1}^m + v_i^m}{2}.$$

One obtains easily the following values of the projections according to the scalar product in $L^2(\mathbb{R})$:

(39)
$$\begin{cases} u_i^{n+1} = - rC_{L,j}^m \, u_j^m + rC_{R,j-1}^m \, u_{j-1}^m + (1 + rC_{L,j}^m - rC_{R,j-1}^m) \, u_i^m \\[2mm] v_i^{m+1} = - rC_{L,j}^m \, v_{1,j}^m + rC_{R,j-1}^m \, v_{2,j-1}^m + (1 + rC_{L,j}^m - rC_{R,j-1}^m) \, v_i^m \\[2mm] p_i^{m+1} = - rC_{L,j}^m \, p_j^m + rC_{R,j-1}^m \, p_{j-1}^m + (1 + rC_{L,j}^m - rC_{R,j-1}^m) \, p_i^m \end{cases}.$$

This method is stable provided
$$1 + rC_{L,j}^m - rC_{R,j-1}^m \geq 0.$$

Therefore stability for both steps is ensured under the condition

$$r \, \mathrm{Max}(C_{R,j-1}^m, -C_{L,j-1}^m, |u_i^n|) \leq \frac{1}{2}.$$

4.6.3 c) <u>Description of the step 3 (antidiffusion).</u> One computes successively

$$\begin{cases} \Delta_j^m = \dfrac{u_{i+1}^m - u_i^m}{2}, \; s_j^m = \text{sign of } \Delta_j^m \; (j = i + \tfrac{1}{2}, \, m = n + \tfrac{1}{2}) \\[4mm] b_j^m = \dfrac{r}{4}(1 - r \dfrac{u_{i+1}^m - u_i^m}{2}) \, ((u_{i+1}^m)^2 - (u_i^m)^2) \\[4mm] a_j^m = s_j^m \, \mathrm{Max}(0, \min(\Delta_{j+1}^m \, s_j^m, \, \Delta_{j-1}^m \, s_j^m, \, |b_j^m|)) \end{cases}$$

and one replaces u_i^m by $u_i^m - a_j^m + a_{j-1}^m$. Similarly for the specific volume and the pressure one sets,

for $w = v$ or p,

$$\begin{cases} \Delta_j^m = \dfrac{w_{i+1}^m - w_i^m}{2}, \quad s_j^m = \text{sign of } \Delta_j^m \\[3em] b_j^m = \dfrac{r}{4}(1\text{-}r\dfrac{u_{i+1}^m + u_i^m}{2})(u_{i+1}^m + u_i^m)(w_{i+1}^m - w_i^m) \\[3em] \text{same formula as above for } a_j^m \end{cases}$$

and one replaces w_i^m by $w_i^m - a_j^m + a_{j\text{-}1}^m$.

4.6.4. The NC2 scheme with conservative projections.

The formulas (39) (L^2 projections of the velocity, the specific volume and the pressure) can be at the origin of defects (that become noteworthy in the case of solid materials submitted to large constraints) : one can observe defects in the conservation of mass, momentum and energy. But projection of $\rho = \dfrac{1}{v}$, instead of v, ensures mass conservation. For the same reason one prefers to project the momentum and the total energy, in place of u and p. To get the new formula (obtained by projection of $\rho = \dfrac{1}{v}$ in place of v) one sets

$$\begin{cases} \alpha = rC_{R,j\text{-}1}^m, \ \beta = rC_{L,j}^m, \ \lambda = 1 - \alpha - \beta \\[1.5em] \dfrac{1}{\rho_i^m} = \dfrac{\alpha}{v_{2,j\text{-}1}^m} + \dfrac{\lambda}{v_i^m} + \dfrac{\beta}{v_{1,j}^m} \end{cases} \quad ;$$

then

$$\begin{cases} \Phi_{1,2} = \rho_i^m(\dfrac{\alpha\lambda}{v_{2,j\text{-}1}^m v_i^m}) \\[1.5em] \Phi_{1,3} = \rho_i^m(\dfrac{\alpha\beta}{v_{2,j\text{-}1}^m v_{1,j}^m}) \\[1.5em] \Phi_{2,3} = \rho_i^m(\dfrac{\lambda\beta}{v_i^m v_{1,j}^m}). \end{cases}$$

The formula giving the specific volume is

(40_a)
$$v_i^{n+1} = \alpha v_{2,j-1}^m + \beta v_{1,j}^m + \lambda v_i^m - \Phi_{1,2}(v_{2,j-1}^m - v_i^m)^2$$

$$- \Phi_{1,3}(v_{2,j-1}^m - v_{i,j}^m)^2 - \Phi_{2,3}(v_{1,j}^m - v_i^m)^2.$$

Projection of momentum and total energy give

(40_b)
$$u_i^{n+1} = \alpha u_{j-1}^m + \beta u_j^m + \lambda u_i^m - \Phi_{1,2}(v_{2,j-1}^m - v_i^m)(u_{j-1}^m - u_i^m)$$

$$- \Phi_{1,3}(v_{2,j-1}^m - v_{1,j}^m)(u_{j-1}^m - u_j^m) - \Phi_{2,3}(v_{1,j}^m - v_i^m)(u_j^m - u_i^m).$$

(40_c)
$$p_i^{n+1} = \alpha p_{j-1}^m + \beta p_j^m + \lambda p_i^m +$$

$$+ \frac{\gamma - 1}{2}[\Phi_{1,2}(u_{j-1}^m - u_i^m)^2 + \Phi_{1,3}(u_{j-1}^m - u_j^m)^2 + \Phi_{2,3}(u_j^m - u_i^m)^2].$$

The remainings are the same as in NC2 (4.6.3) : formulas (40a, b, c) just replace formulas (39).

4.6.5. <u>Numerical results on the shock tube</u>. The problem of the shock tube is the following : air is confined in a tube ; a membrane separates the tube into two parts in which the air is at different pressures and densities. Here are the classical values considered in Sod [1] ($\gamma = 1,4$)

$p = 1$, $\rho = 1$	$p = 0,1$, $\rho = 0,125$
$u = 0$	$u = 0$

At time $t = 0$ the membrane disappears. One has a rarefaction wave travelling to the left, a contact discontinuity and a shock wave travelling to the right, as depicted in the figure below ; they are separated by constant states whose (exact up to the fourth decimal) values are given (c_1 and c_2 are the respective velocities of the contact discontinuity and the shock wave).

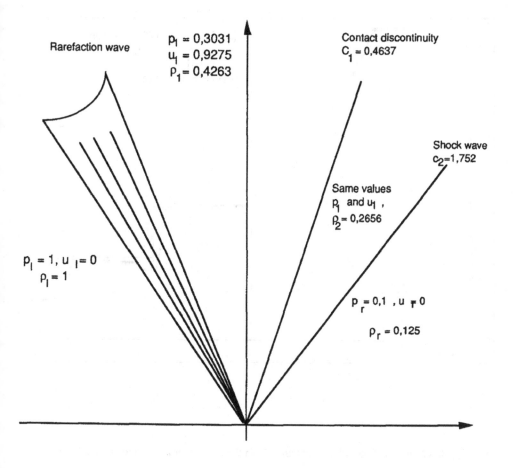

We present below the curves representing the density obtained from the NC1, NC2 and MNC2 (i.e. modified NC2 scheme of 4.6.4) schemes, for the same values of r, h and the same time.

4.6.6 <u>Remark</u>. A splitting of the classical conservative system (25) in which both the convection and the propagation parts are in conservative form has been attempted in Arnaud [1] and compared with 4.6.3 and 4.6.4. In the case of gas dynamics the results are similar, but in the case of heavy media the conservative splitting has to be rejected (see Arnaud [1]). Since v, u, p (and not ρ,u, p) are represented by the same Heaviside function system (30) is more convenient than (25) for the numerical treatment of nonconservative splittings.

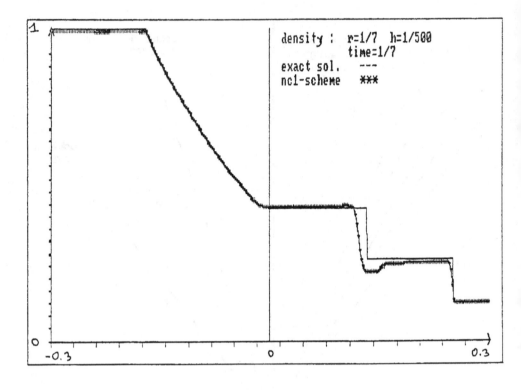

Figure 3. Comparison of the solution obtained from the NC1 scheme with the exact solution.

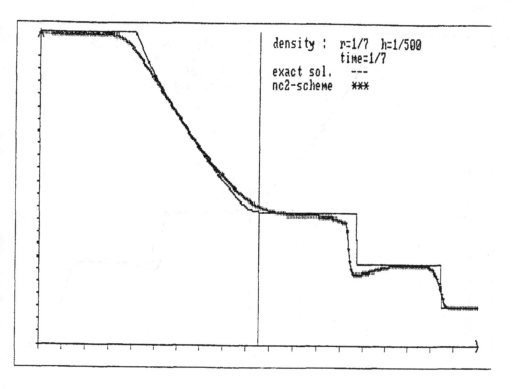

density : r=1/7 h=1/500
           time=1/7
exact sol. ---
nc2-scheme ***

Figure 4. Comparison of the solution obtained from the NC2 scheme with the exact solution.

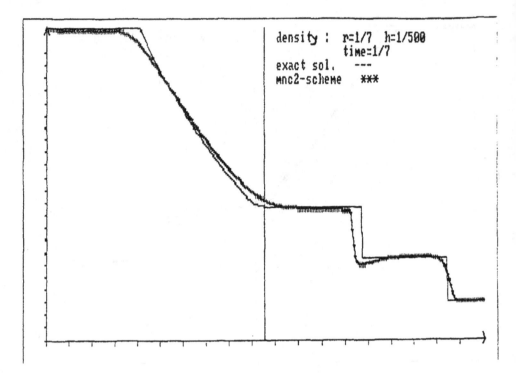

Figure 5. Comparison of the solution obtained from the modified NC2 scheme (i.e. the NC2 scheme with conservative projections) with the exact solution.

The difference between the MNC2 and NC2 schemes is insignificant in the case of gases. It becomes significant in the case of heavy media, see 4.6.9 below, De Luca [1], Noussair [1].

4.6.7 <u>Another method</u>. The nonconservative Lax-Friedrichs method (see problem 3.5.5) can be used to treat the propagation and the convection steps for system (33). It is developped in Berger [1] and gives good results, see figure 6.

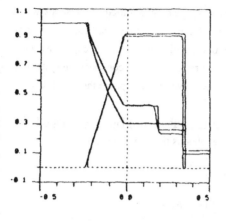

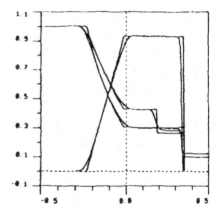

Lax Friedichs' scheme   Godunov NC2's scheme

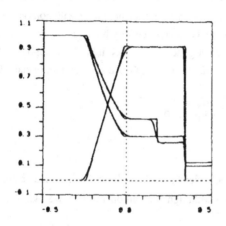

mixing of Godunov and Lax Friedrichs

Figure 6. Numerical solutions of the gas dynamics system (33) using a (nonconservative) Lax Friedrichs type scheme. Comparison with the exact solution ; comparison with the Godunov (NC2) scheme ; very good results are obtained from a mixing of these two schemes.

4.6.8 Conclusion. These schemes do not present oscillations in the neighborhood of the shock wave, of the contact discontinuity and at the foot of the rarefaction wave, as do many classical conservative schemes (see Biagioni [1] App. 3 of Chap. 3 for a comparison with classical schemes). However in gas dynamics there exist already many very good schemes obtained from the conservative system. In the case of solids submitted to large constraints the nonconservative schemes prove their usefulness :

4.6.9. Nonconservative numerical methods in the case of solids submited to large constraints. Using as constitutive equation a Mie Gruneisen equation (32) one obtains the system (30, 30''') i.e.

(41)
$$\begin{cases} v_t + uv_x - vu_x = 0 \\ u_t + uu_x + vp_x = 0 \\ p_t + up_x + [(\gamma + 1)p + F(v)] \, u_x \approx 0 \end{cases}$$

that can be solved similarly as (33) : the mean difference lies in the formulas for the Riemann solver, which are slightly more complicated than those for (33). In this case the numerical values of constant γ or of the coefficents of F (F is usually a polynomial) can be very large, of orders of magnitude 10^6 or 10^{10}. For this reason a very small error on ρ can cause a very large error on p (see (32)), and so most classical conservative methods are unefficient : their oscillations are considerably amplified.

Our nonconservative methods produce very high quality numerical results, see the tests in De Luca [1] in various cases of strong metallic shocks. It has been applied in the 3D axisymmetric case in Arnaud [1] using the form (32') of the Mie Gruneisen equation of state. It gives the system

(42)
$$\begin{cases} v_t + uv_x - vu_x = 0 \\ u_t + uu_x + vp_x = 0 \\ \frac{1}{v} p_t + \frac{u}{v} p_x + [\alpha p + F(v)] \, u_x \approx 0 \end{cases}$$

with a function F and a coefficient α defined from $(30^4, 32')$. For convenience Arnaud [1] has multiplied this equation by v although such an operation is well known to be incorrect ; a correct treatment is given in Berger [1].

§ 4.7 PROBLEM SECTION

Problem 4.7.1. One considers the systems

(S) $\quad \begin{cases} u_t + uu_x \approx 0 \\ \sigma_t + u\sigma_x \approx 0 \end{cases}$
$\qquad\qquad$
(S') $\quad \begin{cases} u_t - \sigma_x \approx 0 \\ \sigma_t - u_x \approx 0 \end{cases}$

Question 1. Compute the travelling wave solutions of system (S) for which $\Delta u \neq 0$ and $\Delta\sigma \neq 0$. Write Godunov's scheme stability condition. Compare with the result from the integration of the system in the cells.

Question 2. Compute the travelling wave solutions of system (S'). Write Godunov's scheme. Stability conditions.

Numerical Question 3. One considers (S) (S') as a convection-propagation splitting of the system

$$(S'') \quad \begin{cases} u_t + uu_x \approx \sigma_x \\ \sigma_t + u\sigma_x \approx u_x \end{cases}$$

i.e. in a first step $n\Delta t \leq t \leq (n + \frac{1}{2})\Delta t$ one solves system (S), obtaining values $(u_i^{n+(\frac{1}{2})}$,

$\sigma_i^{n+(\frac{1}{2})})_{i \in \mathbb{Z}}$ from the known values $(u_i^n, \sigma_i^n)_{i \in \mathbb{Z}}$ and in a second step $(n + \frac{1}{2})\Delta t \leq t \leq (n + 1)\Delta t$

one solves system (S') obtaining values $(u_i^{n+1}, \sigma_i^{n+1})_{i \in \mathbb{Z}}$ from the values $(u_i^{n+\frac{1}{2}}, \sigma_i^{n+\frac{1}{2}})_{i \in \mathbb{Z}}$.

Program system (S'') in this way with the methods of resolution of (S) (S') developped in questions 1 and 2.

Question 4. Solve the Riemann problem for system (S) with one shock wave in which $\Delta\sigma = 0$ and another one in which $\Delta u \neq 0$ (in case $A \neq \frac{1}{2}$). Same questions as 1-3.

Answers. 1) One finds $c = u_1 + \frac{1}{2}\Delta u$: the Riemann problem is solved with only one shock wave ; the stability condition is r. $\underset{i}{\text{Max}} |u_i^n| \leq \frac{1}{2}$. For the scheme one has to distinguish the cases $c > 0$ and $c < 0$.

2) one finds

$$\bar{u} = \frac{u_{i-1}^{n-1} + u_i^{n-1}}{2} + \frac{\sigma_i^{n-1} + \sigma_{i-1}^{n-1}}{2}$$

$$\bar{\sigma} = \frac{\sigma_{i-1}^{n-1} + \sigma_i^{n-1}}{2} + \frac{u_i^{n-1} + u_{i-1}^{n-1}}{2}$$

$$\bar{u} = \frac{u_i^{n-1} + u_{i+1}^{n-1}}{2} + \frac{\sigma_{i+1}^{n-1} + \sigma_i^{n-1}}{2}$$

$$\overline{\overline{\sigma}} = \frac{q_i^{n-1} + \sigma_{i+1}^{n-1}}{2} + \frac{q_{i+1}^{n-1} + u_i^{n-1}}{2}$$

and finally

$$\begin{cases} u_i^n = r\,\overline{u} + r\,\overline{\overline{u}} + (1 - 2r)\,u_i^{n-1} \\[2mm] \sigma_i^n = r\,\overline{\sigma} + r\,\overline{\overline{\sigma}} + (1 - 2r)\,\sigma_i^{n-1} \end{cases}$$

Problem 4.7.2. The purpose of this problem is to give detailed calculations of the NC2 scheme in 4.6.3.

Question 1. Consider the system (convection)

$$(c) \quad \begin{cases} v_t + uv_x \approx 0 \\ u_t + uu_x \approx 0 \\ p_t + up_x \approx 0 \end{cases}$$

in which we assume that the shocks of v, u, p are represented by the same Heaviside function. Show that the Riemann problem is solved as follows

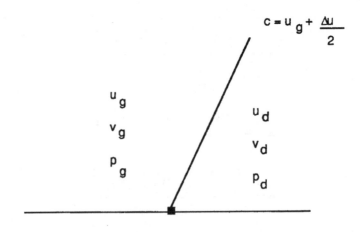

The Godunov scheme can therefore be written in the following way

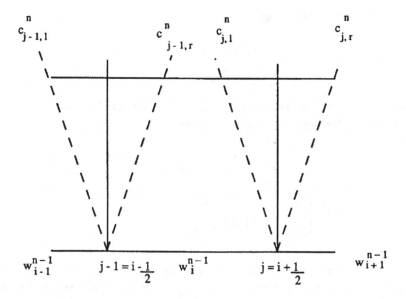

where $w = (u, v, p)$ and where for all j only one of the two velocities $(C_{j,1}^n, C_{j,r}^n)$ does exist ; this is a trick for a convenient statement of the result as follows : we set

$$C_{j,1}^n = \min \left(0, \frac{u_{i-1}^{n-1} + u_i^{n-1}}{2}\right) \le 0 \quad \text{(1 for left)}$$

$$C_{j,r}^n = \max \left(0, \frac{u_{i-1}^{n-1} + u_i^{n-1}}{2}\right) \le 0 \quad \text{(r for right)}.$$

Then set $(r = \dfrac{\Delta t}{\Delta x})$

$$w_i^n = r C_{j-1,r}^n \ w_{i-1}^{n-1} - r C_{j,l}^n \ w_{i+1}^{n-1} + (1 - r C_{j-1,r}^n + r C_{j,1}^n) \ w_i^{n-1}$$

with the stability condition $r \max |u_i^n| < \frac{1}{2}$.

Question 2. Consider the system (propagation)

$$(P) \begin{cases} v_t - v u_x \approx 0 \\ u_t + v p_x \approx 0 \\ p_t + \gamma p\, u_x \approx 0 \end{cases}$$

in which we assume that the shocks of v, u, p are represented by the same Heaviside function. Show that, with the usual notations one obtains the shock relations

$$\begin{cases} c\Delta v = -\dfrac{\Delta u\, \Delta v}{2} - v_1\, \Delta u \\[2mm] c\Delta u = \dfrac{\Delta v\, \Delta p}{2} - v_1\, \Delta p \\[2mm] c\Delta p = \dfrac{\gamma \Delta p\, \Delta u}{2} + \gamma p_1\, \Delta u \end{cases}$$

Remark that a first kind of solution is given by $\Delta u = 0 = \Delta p = c$ (called "contact discontinuity"). A second type of solution is obtained when $\Delta u \neq 0$ and $\Delta p \neq 0$. Check that elimination of c then gives

$$\begin{cases} \Delta u^2 + \Delta p\, \Delta v = 0 \\[2mm] \dfrac{\Delta v}{2} + v_1 + \gamma(\dfrac{1}{2} + \dfrac{p_1}{\Delta p})\, \Delta v = 0 \end{cases}$$

Seek a solution of the Riemann problem in the form

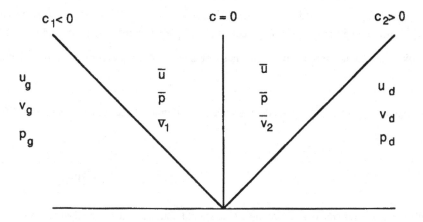

(one has 6 equations for the 6 unknonwns c_1, c_2, $\bar{u}$, $\bar{p}$, $\bar{v}_1$, $\bar{v}_2$).

Setting
$$Z_2 = \sqrt{\gamma \frac{p_d + \bar{p}}{v_d + \bar{v}_2}}$$

show that one has

$$
\begin{cases}
u_d - \bar{u} = \varepsilon\, Z_2(v_d - \bar{v}_2) \quad , \quad \varepsilon = \pm\, 1 \\[2mm]
p_d - \bar{p} = -\,(Z_2)^2\,(v_d - \bar{v}_2) \\[2mm]
c_2 = -\,\varepsilon Z_2\, \dfrac{v_d + \bar{v}_2}{2} \quad .
\end{cases}
$$

The formula for c_2 imposes $\varepsilon = 1$. Similarly for the left shock one obtains, setting $Z_1 = \sqrt{\gamma \dfrac{p_g + \bar{p}}{v_g + \bar{v}_1}}$

$$
\begin{cases}
u_g - \bar{u} = Z_1(v_d - \bar{v}_1) \\[2mm]
p_g - \bar{p} = (Z_1)^2\,(v_g - \bar{v}_1) \\[2mm]
c_1 = -\,Z_1\, \dfrac{v_g + \bar{v}_1}{2}
\end{cases}
$$

The first two equations of these two systems give system (38).

Problem 4.7.3. (continuation of 4.7.2.). Using an approximate Riemann solver for system (38), find w_i^{n+1} from integration of (37) in the cell, and compare with the result obtained in (39) from the L^2 projections. Compare also with the formulas obtained in 40 a, b, c from the L^2 projections of density, momentum and total energy.

Numerical test 4.7.4. Using the double scale method in § 4.2 check that the splitting of (33) given by

(c)
$$\begin{cases} v_t + uv_x = 0 \\ u_t + uu_x = 0 \\ p_t + up_x \approx 0 \end{cases}$$

(p)
$$\begin{cases} v_t - vu_x = 0 \\ u_t + vp_x = 0 \\ p_t + \gamma pu_x \approx 0 \end{cases}$$

gives numerically the correct solution. One should notice that the second equation in (c) is incorrect since it cannot have a shock wave solution ; this suggests that the double scale method gives in fact solutions at different levels of association (the stronger level being interpreted as equality in $\mathcal{G}$ when this is possible). Attempt all other kinds of splittings such as

$$\begin{cases} v_t = 0 \\ u_t + uu_x = 0 \\ p_t + up_x \approx 0 \end{cases}$$

$$\begin{cases} v_t + uv_x - vu_x = 0 \\ u_t + vp_x = 0 \\ p_t + \gamma pu_x \approx 0 \end{cases}$$

Answer : it has been observed that they also lead to the correct solution.

Chapter 5. The case of several constitutive equations.

§5.1 THE AMBIGUITY IN JUMP CONDITIONS FOR THE SYSTEMS OF ELASTICITY AND ELASTOPLASTICITY.

5.1.1. <u>The system of elastoplasticity.</u> We recall the one dimensional system considered in §2.1 :

(1)
$$
\begin{cases}
\rho_t + (\rho u)_x = 0 & \text{balance of mass} \\
(\rho u)_t + (\rho u^2)_x + (p - S)_x = 0 & \text{balance of momentum} \\
(\rho e)_t + [\rho e u + (p - S)u]_x = 0 & \text{balance of energy} \\
S_t + u S_x - k^2(S)u_x \approx 0 & \text{Hooke's law (deviation part) in the elastic case} \\
p \approx \phi(\rho, I, S) & \text{equation of state}
\end{cases}
$$

We recall that ρ = density , u = velocity, $S = S_{11}$ where $(S_{ij})_{1 \le i,j \le 3}$ is the stress deviation tensor, $e =$ density of total energy, $I = e - \frac{1}{2}u^2 =$ density of internal energy. Concerning the constitutive equation we shall distinguish two cases : there is a certain value $S_0 > 0$ (modulus of plasticity) such that if $|S| < S_0$ then the material is in the elastic stage : then $k^2(S)$ is constant (equal to $\frac{4}{3}$ G where G is the shear modulus) ; when $|S| \ge S_0$ then $k^2(S) = 0$ and so the fourth equation in (1) reduces to $S = \pm S_0$: as soon as $|S|$ reaches S_0, it remains constant ; the material is in the plastic stage and is governed by the classical equations of fluid dynamics. For $|S| < S_0$ (elastic stage) the equation of state can be the isotropic part of Hooke's law (2a in chap. 2) or a Mie-Gruneisen equation (32 in chap. 4) ; for $|S| = S_0$ the equation of state is a Mie-Gruneisen equation . According to the general physico-mathematical method in §4.3 the basic laws are stated with equality and the constitutive equations with association. The Mie-Gruneisen equation $p \approx \phi(\rho,I)$ ((32, 32', in chap. 4) is usually adopted as equation of state. Setting $v = \frac{1}{\rho}$ and $\sigma = S - p$ the two first equations in (1) are equivalent to (15' in chap 4) :

(2)
$$
\begin{cases}
v_t + u v_x - v u_x = 0 \\
u_t + u u_x - v \sigma_x = 0
\end{cases}
$$

and one knows that v, u and σ are represented by the same Heaviside function in the case of travelling waves (theorem 4.4.3). The same calculations as in §4.5 give in place of (28) the equation (from the state equation in the form $I \approx \psi(\rho,p)$)

(3) $\qquad - \rho^2 u_x D_1 \psi + \rho(p_t + u p_x)D_2\psi + (p - S)u_x \approx 0.$

Still as in §4.5 setting

(4) $\qquad p \approx \gamma \rho I + G(\rho)$

and

(4') $\qquad F(v) = \frac{1}{v} G' \left(\frac{1}{v}\right) - G\left(\frac{1}{v}\right)$

one obtains from (3) the equation

(3') $\quad p_t + up_x + [(\gamma + 1)p - \gamma S + F(v)]u_x \approx 0$

(another equation similar to 30^4 is obtained from 32'). Therefore in the v, u, p, S variables system (1) appears in the form

(5) $\quad \begin{cases} v_t + uv_x - vu_x = 0 \\ u_t + uu_x + v(p - S)_x = 0 \\ p_t + up_x + [(\gamma + 1)p - \gamma S + F(v)]u_x \approx 0 \\ S_t + uS_x - k^2(S)u_x \approx 0 \end{cases}$

or equivalently in the v, u, σ, S variables

(5') $\quad \begin{cases} v_t + uv_x - vu_x = 0 \\ u_t + uu_x - v\sigma_x = 0 \\ \sigma_t + u\sigma_x - [k^2(S) + S - (\gamma + 1)\sigma + F(v)]u_x \approx 0 \\ S_t + uS_x - k^2(S)u_x \approx 0. \end{cases}$

Since u, v and σ are represented by the same Heaviside function, the ambiguities in jump conditions lie in the terms $k^2(S)u_x$, Su_x and uS_x since one does not know the relative microscopic profiles of u and S on the shock wave.

5.1.2. A resolution of the ambiguity. For the practical purpose of elaborating elastoplastic codes this ambiguity has been solved as follows (Noussaïr [1], Arnaud [1], implicit in Cauret [1]). Often it is easy to build numerical schemes corresponding to the case the unknown functions are represented by the same Heaviside functions (§3.4, §4.6) ; this has been attempted in the case of (1) and (5), (5') (Cauret [1] Noussaïr [1], Arnaud [1]). It has been checked that the results so obtained agree with those from experiments (comparison with classical codes have been published in Cauret [1]).

Of course, in the present case, if a shock wave contains in itself a phase transition : i. e. starting from the elastic stage the material reaches the plastic stage inside the shock wave (and so it is in the plastic stage after the shock) - such shock waves are called elastoplastic shock waves and are very important in practice - then it is clear that S cannot be represented by the same Heaviside function as v, u, p, σ : indeed S ceases to very as soon as |S| reaches the value S_0, see figures 1 and 3 below. So the postulate that the Heaviside functions are the same can only be done at an "infinitesimal level" - to be made clearer below. We postulate

(P) $\quad \begin{vmatrix} \text{for "elementary shock waves" the Heaviside functions of} \\ \\ v, u, p, S, \sigma \text{ are the same.} \end{vmatrix}$

(here "elementary" will mean at the level of a mesh size in a numerical discretization).

From the numerical viewpoint this means that one treats u, v, p, S, σ... in the same way : use of a discretization as the scheme (S_0) in §3.4, or use of a Godunov scheme by imposing in each cell that u, v, p, S, σ... are represented by the same Heaviside function as in §4.6.

Remark that, from the equalities in the two first equation of (5') the Heaviside functions of v,u and σ are the same even on "compound" shock waves (here "compound" means the totality of the microscopic profil of the shock wave).

An attempt to justify (P) from physics was proposed in Colombeau [13]. The real physical phenomenon is a three dimensional one ; with usual notations the two first equations in (5), if stated also in the oy and oz directions would imply that v and σ_{22}, respectively v and σ_{33}, are represented by the same Heaviside function (as well as v and $\sigma = \sigma_{11}$). Since $p = -\frac{1}{3}(\sigma_{11} + \sigma_{22} + \sigma_{33})$ this suggests that p and v are represented by the same Heaviside function : if $\sigma_{ii}(x,t) = \Delta\sigma_{ii} H(x - ct) + \sigma_{iil}$, $i = 1, 2, 3$, (H being the Heaviside function in v) then

$$\sum_{i=1}^{3}\sigma_{ii}(x,t) = (\sum_{i=1}^{3}\Delta\sigma_{ii}) H(x - ct)) + (\sum_{i=1}^{3}\Delta\sigma_{iii}).$$

An argument based on the 3D laws of elasticity is given in Arnaud [1] : we assume the laws of linear elasticity can be applied at the infinitesimal level (i.e. in each cell) ; since the medium is assumed there to be isotropic these laws give

$$\sigma_{11} = (\lambda + 2\mu)u_x$$
$$\sigma_{22} = \lambda u_x$$
$$\sigma_{33} = \lambda u_x .$$

These relations give

$$\sigma_{22} = \sigma_{33} = \frac{v}{1 - v} \sigma_{11}$$

(where $v = \frac{1}{2}\frac{\lambda}{\lambda + \mu}$ is the Poisson coefficient) ; thus the σ_{ii}'s are represented by the same Heaviside function on an infinitesimal level, which justifies (P).

5.1.3. A conservative system of elasticity.

Plohr and Sharp [1] propose a conservative system of elasticity (that however does not apply in the elasto-plastic case). But it applies in the elastic stage i.e. for shock waves in which ISI remains strictly smaller than S_0.

5.1.4. Use of an additional constitutive equation in the elastic case.

In section 2.1 we have mentioned that the isotropic part of Hooke's law and the equation of state are considered as redundant. But since one is confronted with an ambiguity in (5') (namely uS_x, from which on would deduce the one in Su_x) which amounts to the introduction of an additional unknown (the quantity we denote usually by A), then an additional equation is welcome. This additional equation might be ((2a) in chap. 2) $p_t + up_x + a^2u_x \approx 0$, $a > 0$; system (5) (with $k^2(S) = k^2$ since we are concerned here with the elastic case) becomes

$$(6) \quad \begin{cases} v_t + uv_x - vu_x = 0 \\ u_t + uu_x - v\sigma_x = 0 \\ \sigma_t + u\sigma_x - (k^2 + a^2)u_x \approx 0 \\ p_t + up_x + a^2u_x \approx 0 \\ p_t + up_x + (p - \gamma\sigma + F(v))u_x \approx 0 \end{cases}$$

i.e.

$$(6') \quad \begin{cases} v_t + uv_x - vu_x = 0 \\ u_t + uu_x - v\sigma_x = 0 \\ \sigma_t + u\sigma_x - (k^2 + a^2)u_x \approx 0 \\ p_t + up_x + a^2u_x \approx 0 \\ (p - \gamma\sigma + F(v) - a^2)u_x \approx 0. \end{cases}$$

Representing v, u, σ by a Heaviside function H and p by a Heaviside function K, the last equation in (6') gives

$$(7) \quad [\Delta pK + p_\ell - \gamma\Delta\sigma H - \gamma\sigma_\ell + F(\Delta vH + v_\ell) - a^2] H' \approx 0.$$

Setting $HK' \approx A\delta$ (this follows from the fourth equation in (6')) and so $KH' \approx (1 - A)\delta$,

(7)   gives by integration

$$(7') \quad \Delta p(1 - A) + p_\ell - \gamma\frac{\Delta\sigma}{2} - \gamma\sigma_\ell - a^2 + \int_{v_\ell}^{\Delta u + v_\ell} F(\lambda)d\lambda = 0$$

which gives a value of A and so resolves the ambiguity. The calculation of A from (7') can be avoided : the last equation in (6') gives pu_x as a function of σ and v and then $up_x = (up)_x - pu_x$ appears in the form of a conservative term and a function of σ and v, to be put in the fourth equation of (6').

This way of resolving the ambiguity has not yet been compared with those in 5.1.2 and 5.1.3. It is exposed here only to show that one can resolve the ambiguity form additional equations.

§5.2. CALCULATIONS OF ELASTOPLASTIC SHOCK WAVES AND ELASTIC PRECURSORS.

The aim of this section is to show on a simplified model of elastoplasticity that the method in 5.1.2 resolves the ambiguity. For simplicity in explicit calculations we consider only shocks in which the density varies slightly in the neighborhood of a fixed value ρ_0 (this assumption is not always justified physically : in strong metallic shocks the relative variation in density can reach 0, 1 or more). Then, as explained in 4.3.5, the first two equations in (5') may be replaced by the equation

$$\rho_0(u_t + uu_x) - \sigma_x \approx 0$$

together with the assumption that u and σ are represented by the same Heaviside function. Setting $v_0 = \dfrac{1}{\rho_0}$ system (5') is

(8)
$$\begin{cases} u_t + uu_x - v_0\sigma_x \approx 0 \\ \sigma_t + u\sigma_x - [-a\sigma + S + b + k^2(S)]\, u_x \approx 0 \\ S_t + uS_x - k^2(S)u_x \approx 0 \end{cases}$$

where

$$k^2(S) = k^2 \text{ if } |S| \leq S_0 \text{ and } k^2(S) = 0 \text{ if } |S| = S_0$$

k^2, a, b, v_0 given real numbers. Further we assume that u and σ are represented by the same Heaviside function (from the strong statement of the two first equations in (5')) and we assume (P) : u, σ and S are represented by the same Heaviside functions in "infinitesimal shocks" (as above i.e, at the size level of a mesh size).

5.2.1. Elastoplastic shock waves. There are shock waves throughout which the material passes from the elastic stage to the plastic stage. For such a shock wave one can imagine that, at an infinitesimal level of description, |S| increases and reaches the critical value S_0, and then ceases to vary while u and σ go on varying. Thus, unless the plastic stage would be null, S cannot be represented by the same Heaviside function as u and σ, see figure 1 below.

For simplicity in calculations let us consider only a shock wave in which the respective values of (u, p, S) are (0, 0, 0) on the right hand side and $(u_\ell, p_\ell, -S_0)$ on the left hand side (shock wave in a target initially at rest, produced by a projectile coming from the left hand side). The microscopic profile of the shock wave under consideration is as follows :

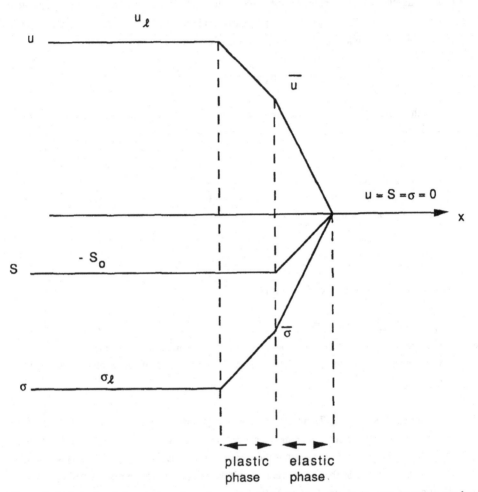

sense of velocity of the shock

Figure 1. Infinitesimal description of an elasto-plastic shock wave (shock in a target at rest produced by a projectile coming from the left hand side). Compare with figure 3 that represents the result of a numerical test in the same situation.

In a simplified case (obtained by dropping a few terms) the second equation in (8) is

(8')
$$\sigma_t + u\sigma_x - d^2 u_x \approx 0 \; ;$$

one shows easily that as soon as one remains in the elastic case then the Heaviside functions of u, σ and S are the same on a shock. Indeed one considers then the simplified system

(9)
$$\begin{cases} u_t + u u_x - v_0 \sigma_x \approx 0 \\ \sigma_t + u\sigma_x - d^2 u_x \approx 0 \\ S_t + u S_x - k^2 u_x \approx 0. \end{cases}$$

Setting, in an "infinitesimal shock" (for instance in a cell)

$$w(x,t) = \Delta w H(x - ct) + w_\ell$$

then (9) gives

(10)
$$\begin{cases} c - \dfrac{\Delta u}{2} = u_\ell - v_0 \dfrac{\Delta \sigma}{\Delta u} \\[2mm] c - \dfrac{\Delta u}{2} = u_\ell - d^2 \dfrac{\Delta u}{\Delta \sigma} \\[2mm] c - \dfrac{\Delta u}{2} = u_\ell - k^2 \dfrac{\Delta u}{\Delta S}. \end{cases}$$

Elimination of c gives

(10')
$$\begin{cases} \dfrac{\Delta u}{\Delta \sigma} = \theta \dfrac{\sqrt{v_0}}{d} \\[3mm] \dfrac{\Delta u}{\Delta S} = \theta \dfrac{d\sqrt{v_0}}{k^2}. \end{cases} \quad \text{with } \theta = +1 \text{ or } \theta = -1$$

In the shocks under consideration we admit that θ keeps the same value, when the shock is a superposition of a certain number of "elementary shocks" (for instance those in a cell).

Consider a situation of a (finite) number of successive shocks in which the physical variables are represented by the same Heaviside function and have proportional jumps. Then in the compound shock the physical variables are still represented by the same Heaviside functions.

From (10') this leads us to represent u, σ, S by the same Heaviside functions as long as one remains in the elastic case (note that this is a consequence of (P) for "infinitesimal" or "elementary" shocks and that this is also valid only for system (9)).

This also applies in the plastic stage for u and σ (then $S \equiv \pm S_0$). Therefore if one considers the elastoplastic system

$$
(11) \quad \begin{cases} u_t + uu_x - v_0\sigma_x \approx 0 \\ \sigma_t + u\sigma_x + (\ell(S)\sigma + m(S))u_x \approx 0 \\ S_t + uS_x - k^2(S)u_x \approx 0. \end{cases}
$$

with

$$
(11') \quad \begin{cases} \ell(S) = 0 \ , \quad m(S) = -d^2 \ , \quad k^2(S) = k^2 \quad \text{if } |S| < S_0 \\ \ell(S) = \alpha \ , \quad m(S) = \beta \ , \quad k^2(S) = 0 \quad \text{if } |S| = S_0, \end{cases}
$$

then one represents analytically solutions u, S, σ in figure 1 in the form

$$
(12) \quad \begin{cases} u(x,t) = (u_\ell - \bar{u}) \, H(-x + ct) + \bar{u}K(-x + ct) \\ \sigma(x,t) = (\sigma_\ell - \bar{\sigma}) \, H(-x + ct) + \bar{\sigma}K(-x + ct) \\ S(x,t) = -S_0 K(-x + ct) \end{cases}
$$

where H and K are two (non overlapping : HK' = 0) Heaviside generalized functions, and where c is the velocity of the elastoplastic shock wave.

$\dfrac{u(x,t)}{u_\ell}$ and $\dfrac{\sigma(x,t)}{\sigma_\ell}$ are equal since they are (in the variable x - ct) the Heaviside functions of u and σ on the global elastoplastic shock (from (6c) these Heaviside functions are equal).

This gives

$$
(1 - \frac{\bar{u}}{u_\ell}) H + \frac{\bar{u}K}{u_\ell} = (1 - \frac{\bar{\sigma}}{\sigma_\ell}) H + \frac{\bar{\sigma}}{\sigma_\ell} K
$$

i.e. (from the disjunction of H and K)

$$
(13) \quad \frac{\bar{u}}{u_\ell} = \frac{\bar{\sigma}}{\sigma_\ell} \ .
$$

Putting (12) into (11) one obtains

$$
(14) \quad cu_\ell - \frac{(u_\ell - \bar{u})^2}{2} - \frac{\bar{u}^2}{2} + v_0\sigma_\ell = 0
$$

$$
(15) \quad c\sigma_\ell - (1 + \alpha)\frac{(u_\ell - \bar{u})(\sigma_\ell - \bar{\sigma})}{2} - \frac{\bar{u}\,\bar{\sigma}}{2} - \beta \, u_\ell + d^2\bar{u} = 0
$$

$$
(16) \quad cS_0 - \frac{\bar{u}S_0}{2} - k^2\bar{u} = 0.
$$

(13), (14), (15), (16) are a system of four equations for the 6 unknown numbers c, $\bar{u}$, $\bar{\sigma}$, u_ℓ, σ_ℓ. In the case only elastoplastic shock waves (with an elastic-plastic transition inside them) are concerned this leads to the following resolution of the Riemann problem : each shock wave gives four equations and one has eight unknown numbers : c, c', $\bar{u}$, $\bar{\sigma}$, $\bar{u}\,'$, $\bar{\sigma}'$, u_1, σ_1

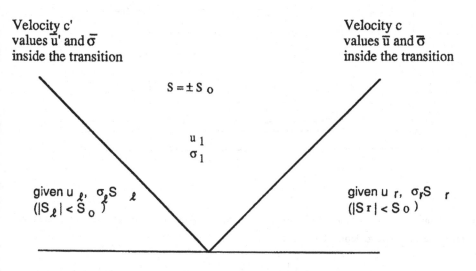

Velocity c'
values $\bar{u}'$ and $\bar{\sigma}$
inside the transition

Velocity c
values $\bar{u}$ and $\bar{\sigma}$
inside the transition

$S = \pm S_o$

u_1
σ_1

given u_ℓ, σ_ℓ, S_ℓ
$(|S_\ell| < S_o)$

given u_r, σ_r, S_r
$(|S_r| < S_o)$

Figure 2. Schematic representation of the solution of the Riemann problem for system (11) when only elastoplastic shock waves are concerned. This Riemann problem can represent the collision of a projectile on a target, see figure 3.

Using a numerical scheme analogous to the (S_0) scheme in §3.4 one obtains the following numerical result (in complete agreement with fig. 1 and 2) :

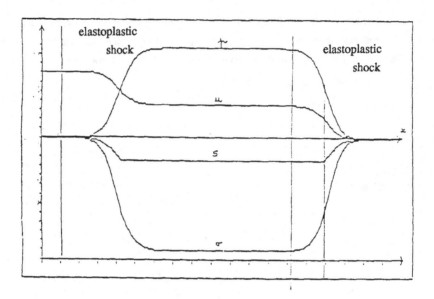

<u>Figure 3.</u> A numerical solution of the Riemann problem for system (11) ; on the right hand side one recognizes the shock wave depicted in fig. 1.

5.2.2 <u>Elastic precursors.</u> In the case the difference of velocities $|u_r - u_\ell|$ (in fig. 2) is smaller or in the case $|S_0|$ is smaller then one observes a different result : instead of an elastoplastic shock wave one observes two shock waves :

    - one in the elastic stage in which the value of $|S|$ is raised to S_0 (or starts from S_0 and reaches a smaller value) ; this shock wave is called an elastic precursor because in the target, fig. 4, it preceeds the second shock wave. These elastic precursors have usually a rather weak amplitude and are not very destructive.

- one in the plastic stage, in which |S| remains constant equal to S_0, called a plastic shock wave. These plastic shock waves are usually very destructive.

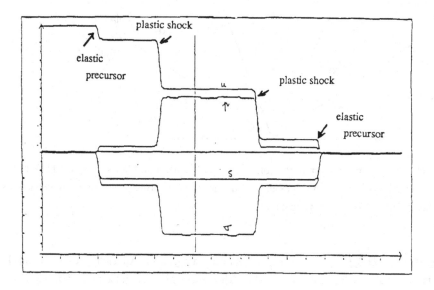

Figure 4. A numerical solution of the Riemann problem for (11), in a case in which there appear elastic precursors (same numerical scheme as in fig. 3, but different initial conditions : smaller velocity of the projectile or smaller value of S_0).

One can easily do explicit calculations as this is done in 5.2.1 for elastoplastic shock waves. For the elastic precursors one uses the elastic form (9) of (11) and for the plastic waves one uses the plastic form of (11) :

$$\begin{cases} u_t + uu_x - v_0\sigma_x \approx 0 \\ \sigma_t + u\sigma_x + (\alpha\sigma + \beta)u_x \approx 0. \end{cases}$$

The schematic depiction of the Riemann problem is as follows :

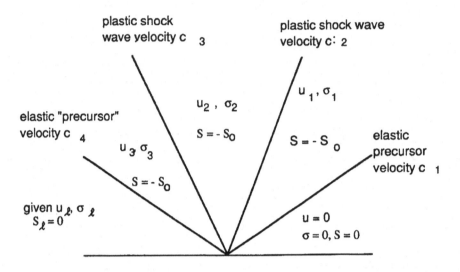

Figure 5. Schematic representation of the solution of the Riemann problem for system (11), in the case of fig. 4. One has 10 unknown values and 10 equations (each elastic precursor gives 3 equations and each plastic shock wave gives 2 equations).

§5.3. <u>NUMERICAL RESULTS</u>.Various numerical results are given in Appendix 4 of Biagioni [1] ; other results are exposed in figure 6   to 10 below. They are obtained from the method introduced in Problems 3.5.5 and 4.7.5, see a detailed study in Berger [1] (use of Godunov scheme, Lax-Friedrichs scheme and a mixing of both).

Figures 6, 7, 8 represent numerical solutions of (constant density) system

$$(S_1) \quad \begin{cases} u_t + uu_x + (p - S)_x = 0 \\ p_t + up_x + (2p - S) u_x = 0 \\ S_t + uS_x - \mu(S) u_x = 0 \end{cases}$$

Figures 9, 10 compare numerical solutions of (S_1) with the (variable density) system

$$(S_2) \quad \begin{cases} v_t + uv_x - vu_x = 0 \\ u_t + uu_x + v(p - S)_x = 0 \\ p_t + up_x + (2p - S)\, u_x = 0 \\ S_t + uS_x - \mu(S)\, u_x = 0 \end{cases}$$

where $\mu(S) = \begin{cases} 2 & \text{if } |S| < S_o \\ 0 & \text{if } |S| = S_o \end{cases}$, $S_o = 0, 1$ and for the initial conditions

$$u_g = 0, 5 \qquad u_d = 0$$
$$p_g = 0, 3 \qquad p_d = 0, 3$$
$$S_g = 0 \qquad S_d = 0$$
$$\rho_g = 1 \qquad \rho_d = 1.$$

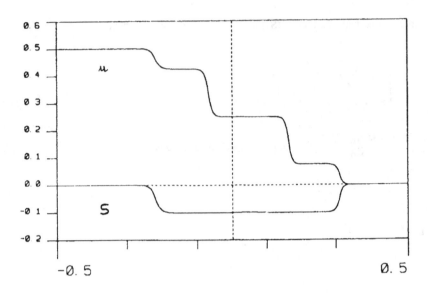

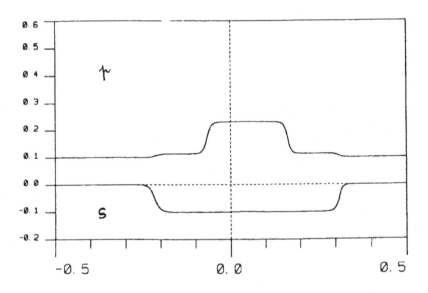

Figure 6. System (S$_1$) with the initial conditions u$_g$ = 0, 5 , p$_g$ = 0, 1, S$_g$ = 0, u$_d$ = 0 , p$_d$ = 0, 1, S$_d$ = 0 and with μ(S) = 2 if ISI < S$_o$, μ(S) = 0 if ISI = S$_o$, with S$_o$ = 0, 1. One observes elastic precursors.

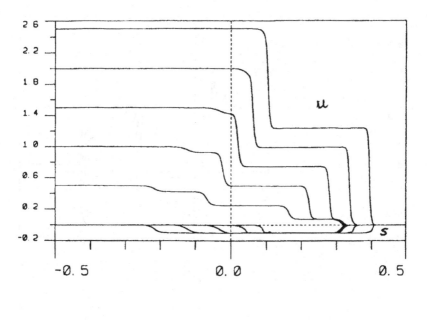

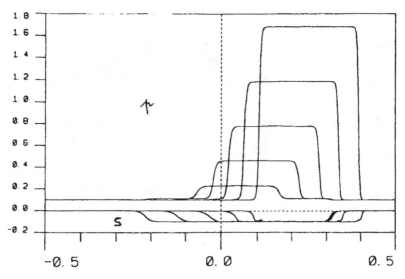

Figure 7. Relatively to fig 6 one increases the velocity u_g of the projectile : the velocities of the precursor get closer to those of the hydrodynamic waves and then the precursors get confounded with the hydrodynamic waves, forming elastoplastic shockwaves.

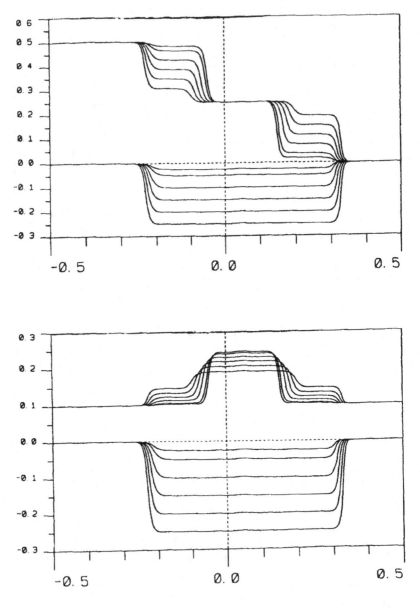

Figure 8. Relatively to fig 6 one diminishes the value of S_0 : the relative velocities of the precursors and the hydrodynamic waves do not vary significantly, but the amplitude of the precursors tend to zero.

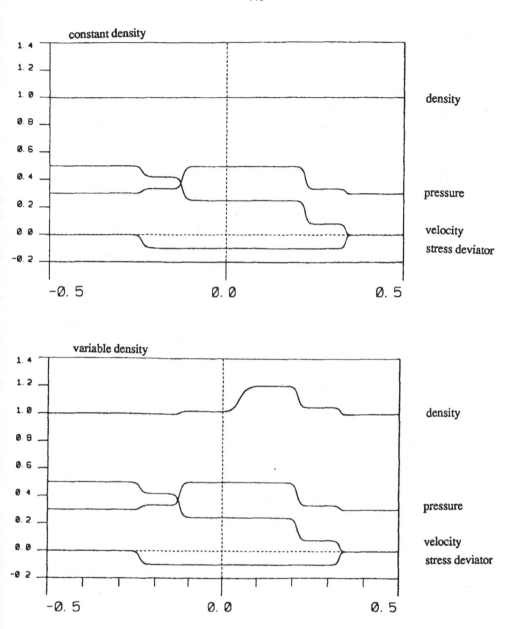

Figure 9 : Comparison of (S$_1$) (on top) and (S$_2$) (bottom) : one observes that p, u and S are not significantly affected by the presence of $\rho = \frac{1}{v}$ in (S$_2$) : see fig. 10.

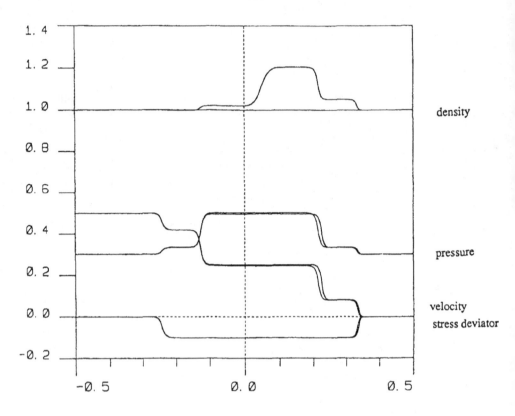

Figure 10 : Superposition of the curves in fig. 9.

§5.4. MULTIFLUID FLOWS

5.4.1. Statement of the equations. Let us consider a mixture of two fluids ; let p denotes the pressure of the mixture, ρ_i, u_i (i = 1,2) the respective density and velocity of fluid number i (at each point (x,t)) ; let α denote the volumic proportion of the fluid number 1 in the mixture ($0 \leq \alpha(x,t) \leq 1$). Then commonly adopted equations are system (8) in chapter (2); we drop the energy equations for simplification :

(17)
$$
\begin{cases}
(\alpha\rho_1)_t + (\alpha\rho_1 u_1)_x = 0 & \text{balance of mass for fluid 1} \\
((1-\alpha)\rho_2)_t + ((1-\alpha)\rho_2 u_2)_x = 0 & \text{balance of mass for fluid 2} \\
(\alpha\rho_1 u_1)_t + (\alpha\rho_1 u_1^2)_x + \alpha p_x = 0 & \text{balance of momentum for fluid 1} \\
((1-\alpha)\rho_2 u_2)_t + ((1-\alpha)\rho_2 u_2^2)_x + (1-\alpha)p_x = 0 & \text{balance of momentum for fluid 2} \\
\rho_1 \approx \rho_1(p) , \rho_2 \approx \rho_2(p) & \text{equations of state of fluids 1 and 2}
\end{cases}
$$

The terms αp_x and $(1-\alpha)p_x$ show multiplications of distributions of the form $Y \cdot \delta$ in the case of shock waves.

Our analysis in §4.3 suggests to state the four first equations with equality in $\mathcal{G}$: this means that we assume their relevance in space time volumes smaller than the "width of shock waves". In order to transform these equations similarly as those of one fluid flow in §4.5 we set

(18)
$$
\begin{cases}
r_1(x,t) = \alpha(x,t)\rho_1(x,t) \\
r_2(x,t) = (1 - \alpha(x,t)) \rho_2(x,t).
\end{cases}
$$

Then the first four equations in (17) become :

(19)
$$
\begin{cases}
(r_1)_t + (r_1 u_1)_x = 0 \\
(r_2)_t + (r_2 u_2)_x = 0 \\
(r_1 u_1)_t + (r_1 u_1^2)_x + \alpha p_x = 0 \\
(r_2 u_2)_t + (r_2 u_2^2)_x + (1 - \alpha)p_x = 0.
\end{cases}
$$

The first and the third equations give
$$r_1(u_1)_t + r_1 u_1(u_1)_x + \alpha p_x = 0.$$

Assuming α is nowhere 0 (i.e. that our mixture is really a mixture everywhere) we can set

(20)
$$v_1 = \frac{1}{r_1} = \frac{1}{\alpha\rho_1} , \quad v_2 = \frac{1}{r_2} = \frac{1}{(1-\alpha)\rho_2} .$$

Immediate calculations (as in 4.4.2) transform (19) into the equivalent system

$$(21) \quad \begin{cases} (v_1)_t + u_1(v_1)_x - v_1(u_1)_x = 0 \\ (v_2)_t + u_2(v_2)_x - v_2(u_2)_x = 0 \\ (u_1)_t + u_1(u_1)_x + \alpha v_1 p_x = 0 \\ (u_2)_t + u_2(u_2)_x + (1 - \alpha)v_2 p_x = 0 \end{cases}$$

5.4.2. Jump conditions.

We seek travelling wave solutions of (21) in the form

$$(22) \quad \begin{cases} u_i(x,t) = \Delta u_i H_i(x - ct) + u_{i\ell} \quad i = 1,2 \\ v_i(x,t) = \Delta v_i M_i(x - ct) + v_{i\ell} \quad i = 1,2 \\ p(x,t) = \Delta p K(x - ct) + p_\ell \\ \alpha(x,t) = \Delta \alpha L(x - ct) + \alpha_\ell \end{cases}$$

where c, Δu_i, Δv_i, Δp, $\Delta \alpha$, $u_{i\ell}$, $v_{i\ell}$, $p_{i\ell}$, α_ℓ are real numbers and where H_i, M_i, K, L, $\in \mathcal{G}(\mathbb{R})$ are Heaviside generalized functions.

From calculations in 4.4.2 (6) (14') (16') the first equation in (21) gives (if $\Delta v_1 \neq 0$)

$$M_1' - \frac{H_1'}{a+H_1} \quad M_1 - \frac{a\,H_1'}{a+H_1} = 0 \quad \text{if } a = \frac{v_{1\ell}}{\Delta v_1}$$

from which one deduces $M_1 = H_1$. Similarly the second equation in (21) yields $M_2 = H_2$. Note that the two first equations in (21) are respectively equivalent to

$$(23) \quad \begin{cases} M_1 = H_1 \text{ and } c - u_{1\ell} = - v_{1\ell} \dfrac{\Delta u_1}{\Delta v_1} \\ M_2 = H_2 \text{ and } c - u_{2\ell} = - v_{2\ell} \dfrac{\Delta u_2}{\Delta v_2}. \end{cases}$$

Now let us consider the third equation in (21). Using (22), (23) and assuming $a + H_1 \neq 0$ one obtains

$$(24) \quad (\Delta \alpha\, L + \alpha_\ell) K' = - \frac{(\Delta u_1)^2}{\Delta p \Delta v_1} H_1'.$$

Similarly the fourth equation in (21) gives

$$(24') \quad (1 - \Delta \alpha L - \alpha_\ell) K' = - \frac{(\Delta u_2)^2}{\Delta p \Delta v_2} H_2'.$$

Adding (24) and (24') gives the relation

$$K' = -\frac{1}{\Delta p}\left(\frac{(\Delta u_1)^2}{\Delta v_1}H_1' + \frac{(\Delta u_2)^2}{\Delta v_2}H_2'\right)$$

i.e.

(25)
$$K = -\frac{1}{\Delta p}\left(\frac{(\Delta u_1)^2}{\Delta v_1}H_1 + \frac{(\Delta u_2)^2}{\Delta v_2}H_2\right)$$

which implies the jump relation

(26)
$$\frac{(\Delta u_1)^2}{\Delta v_1} + \frac{(\Delta u_2)^2}{\Delta v_2} = -\Delta p.$$

Finally one has obtained (assuming α does not take the value 0)

<u>Theorem</u>. (22) is solution of (21) if and only if

$$(27)\begin{cases} M_1 = H_1 \text{ and } c - u_{1\ell} = -v_{1\ell}\dfrac{\Delta u_1}{\Delta v_1} \\[2mm] M_2 = H_2 \text{ and } c - u_{2\ell} = -v_{2\ell}\dfrac{\Delta u_2}{\Delta v_2} \\[2mm] K = -\dfrac{1}{\Delta p}\left(\dfrac{(\Delta u_1)^2}{\Delta v_1}\right)H_1 + \dfrac{(\Delta u_2)^2}{\Delta v_2}H_2),\ \text{which implies the relation (26)} \\[2mm] (\Delta\alpha L + \alpha_\ell)K' = -\dfrac{(\Delta u_1)^2}{\Delta p \Delta v_1}H_1',\ \text{which gives the relation}\ \begin{cases} LK' \approx A\delta \\[1mm] A\Delta\alpha + \alpha_\ell = -\dfrac{(\Delta u_1)^2}{\Delta p \Delta v_1}. \end{cases} \end{cases}$$

The product LK' yields an ambiguity and so one has only three well defined jump relations, i.e. five jump relations for system (17) : of course these are the same as the Rankine Hugoniot jump conditions obtained by adding the two balance of momentum equations in (17). To resolve the ambiguity we need to know the relative microscopic profile of α and p on a shock, as already obvious from (17).

5.4.3 <u>Remark</u>. From the last equation in (27) $K = L = H_1$ is impossible unless $\Delta\alpha = 0$ (since $HH' \neq aH'$, $a \in \mathbb{R}$: this would imply $a = \frac{1}{2}$, $H^2 = H$ by integration, and then $H^n = H$ which is proved to be impossible in §3.1), but $K = L$ is possible.

An additional piece of information is needed to resolve the ambiguity. Here we present various proposals : they rely on the relative behavior of the two fluids.

5.4.4. <u>Mixture of two gases that are assumed to satisfy a law p.V = constant inside a shock (p = pressure, V = volume)</u>. Let us try to determine the microscopic profile of α as a function of the one of p.

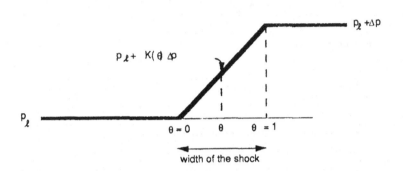

<u>Figure 6</u>: the jump in pressure in a shock wave.

Let $X(\theta)$ and $Y(\theta)$ be the volumes of the 1st and 2nd fluid respectively when $p = p_\ell + K(\theta)\Delta p$, if the initial volumes are α_ℓ and $1 - \alpha_\ell$ respectively. From our assumption $pV = $ constant one has

$$(28) \quad \begin{cases} X(\theta).\,(p_\ell + K(\theta)\Delta p) = \alpha_\ell p_\ell \\ Y(\theta).\,(p_\ell + K(\theta)\,\Delta p) = (1 - \alpha_\ell)p_\ell. \end{cases}$$

Let $\alpha(\theta)$ be the volumic proportion of fluid (1) when $p = p_\ell + K(\theta)\Delta p$. Then

$$\alpha(\theta) = \frac{X(\theta)}{X(\theta) + Y(\theta)} = \frac{\alpha_\ell p_\ell}{\alpha_\ell p_\ell + (1 - \alpha_\ell)p_\ell} = \alpha_\ell.$$

This means that α is constant on a shock. Then in (27) one has $\Delta\alpha = 0$ and the relation $\alpha_\ell = -\frac{(\Delta u_1)^2}{\Delta p\Delta v_1}$. The term αp_x is no longer a product of distributions.

5.4.5 <u>Mixture of a liquid and a gas</u> (idea communicated to us by L. Sainsaulieu [1]).
We assume that the liquid (= fluid number 1) is not compressible inside the shock and we drop its state law. Assuming α is nowhere zero the first and third equation in (17) give

$$\begin{cases} \alpha_t + (\alpha u_1)_x = 0 \\ (u_1)_t + u_1\,(u_1)_x + \dfrac{1}{\rho_1}\,p_x = 0 \end{cases}$$

thus the nonconservative term $\propto p_x$ disappears and our system becomes conservative.

5.4.6 A general resolution of the ambiguity in case of two very different fluids (liquid and a gas for instance). Let us assume the constitutive equation of fluid 1 is exactly valid inside a shock, i-e:

$\rho_1 = \rho_1(p)$ in (17). From (20) state it as $\alpha\, v_1 = \varphi(p)$. From (22)

$$(\Delta\alpha\, L(\xi) + \alpha_\ell)\,(\Delta v_1\, H_1(\xi) + v_{1\ell}) = \varphi\,(\Delta p\, K(\xi) + p_\ell).$$

Multiplying by $\Delta p\, K'(\xi)$ and using (27)

$$-\frac{(\Delta u_1)^2}{\Delta v_1}\, H_1'\,(\xi)\,(\Delta v_1\, H_1\,(\xi) + v_{1\ell}) = \varphi\,(\Delta p\, K(\xi) + p_\ell)\,\Delta p\, K'(\xi).$$

Integration yields

$$-\frac{(\Delta u_1)^2}{(\Delta v_1)^2}\,\frac{1}{2}\,(\Delta v_1\, H_1\,(\xi) + v_{1\ell})^2 = \int_0^{\Delta p K(\xi) + p_\ell} \varphi\,(\lambda)\, d\lambda + \text{constante}$$

which gives ($\xi = +\infty$ and $-\infty$)

$$(29)\quad -(\Delta u_1)^2\left(\frac{1}{2} + \frac{v_{1\ell}}{\Delta v_1}\right) = \int_{p_\ell}^{\Delta p + p_\ell} \varphi\,(\lambda)\, d\lambda.$$

(29) replaces the (undefined as long as it depends on A) last relation in (27). It solves the ambiguity (four jump relations for four equations). Then the last equation in (27) gives the corresponding value for A.

5.4.7 Resolution of the ambiguity : a practical assumption for numerical discretization. The resolution of the ambiguity should be done for the complete system in Stewart-Wendroff [1] including also 2 energy equations, and validated from comparison with experimental data : comparison with classical schemes (Liles-Reed [1]) shows that (at least roughly) these classical schemes give results corresponding to the postulate that all Heaviside functions in (22) are the same at the infinitesimal level of a mesh size, as in §5.2 concerning elastoplasticity. In the sequel we adopt this simplifying assumption, although it contradicts 5.4.3 when applied to the variables in (22).

This assumption can be justified as follows if one assumes $\Delta\alpha$ small relatively to α_ℓ and if we assume that the Heaviside functions K and L are monotonic (from 0 to 1) as this is natural. Then $0 < A < 1$ (obvious from fig. 4 in chap. 2).

The formula

$$A \, \Delta\alpha + \alpha_\ell = - \frac{(\Delta u_1)^2}{\Delta p \, \Delta v_1}$$

can be stated approximately as $\alpha_\ell = - \dfrac{(\Delta u_1)^2}{\Delta p \, \Delta v_1}$, which resolves the ambiguity if $\Delta\alpha$ is neglected relatively to α_ℓ. But one may assume that this holds at the infinitesimal level of a meshsize since in practice a certain number of meshes are involved in a shock (for such a complicated system of 6 equations). Then, at a meshsize level, the ambiguity disappears under the above approximation. Since it is convenient to have all Heaviside functions identical then we adopt the above postulate, but of course we do not pretend to attain in this way even an approximation of the case α takes values close to 0.

5.4.8 Remark in the case of the full system ChapII, (7) involving energy equations. The energy equation of the first fluid is

$$(\alpha \, \rho_1 \, e_1)_t + (\alpha \, \rho_1 \, e_1 \, u_1 + \alpha \, p \, u_1)_x + p \, \alpha_t = 0$$

and the ambiguity lies in the term $p\alpha_t$. From (22)

$$p \, \alpha_t = (\Delta p \, K + p_\ell)(-c \, \Delta \, \alpha \, L') = -c \, \Delta \, \alpha \, (\Delta \, p \, K \, L' + p_\ell \, L').$$

Thus

$$p \, \alpha_t \approx -c \, \Delta \, \alpha \, ((1-A)\Delta \, p + p_\ell).$$

Since $0 \le A \le 1$ the ambiguity is insignificant when $|\Delta p|$ is small relatively to p_ℓ. This provides an easy practical solution for numerical purposes. Without this approximation it is clear that as in 5.4.6 the statement of the state law of fluid 1 with equality solves the ambiguity. Let us consider the particular case $\rho_1 = $ constant and one drops the state law of the first fluid. One has the equations

$$\begin{cases} \alpha_t + (\alpha u_1)_x = 0 \\ (u_1)_t + u_1 \, (u_1)_x + \dfrac{1}{\rho_1} \, p_x = 0 \end{cases}$$

A calculation analog to the one in 4.4.2 permits to obtain that the Heaviside functions inside $\frac{1}{\alpha}$ and u_1 are the same, and to express the Heaviside function of p in term of it. Then the ambiguity $p\alpha_t$ in the energy equation is solved.

§5.5. NUMERICAL RESULTS. We give three numerical solutions of systems (17) under the method in 5.4.7 and a splitting in two steps (see Berger [1]). In the case of only one fluid similar results are given in figure 6 of chap 4. One uses Godunov's scheme, Lax-Friedrich's scheme and a mixing of both, see Berger [1].

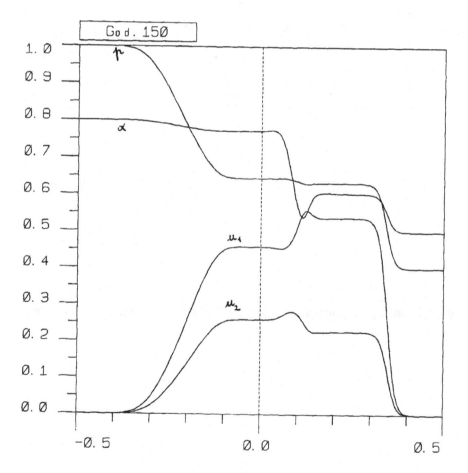

Figure 7. Solution of (17) from Godunov's scheme (applied on each system of the splitting).

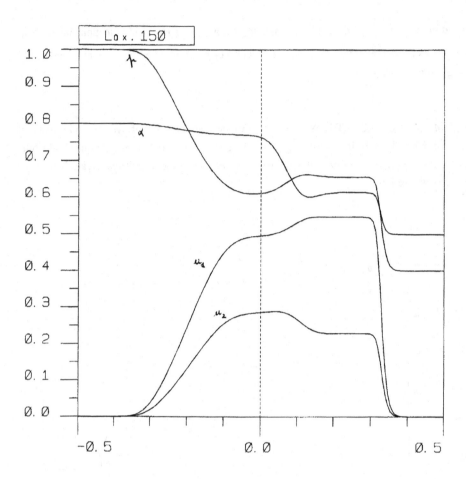

Figure 8. Solution of (17) from a nonconservative Lax-Friedrichs scheme (on each system of the splitting).

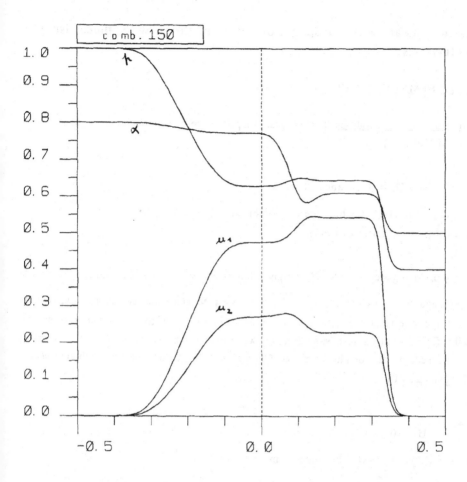

Figure 9. Solution of (17) from a mixing of Godunov and Lax-Friedrichs.

Chapter 6. Linear wave propagation in a medium with piecewise C^∞ characteristics.

§ 6.1 A MATHEMATICAL STUDY.

6.1.1 Equations of linear acoustics . In the Lagrangian picture the acoustic waves are described by the system ((15) of chap. 2)

$$(1)\begin{cases} \rho_t + \rho_0 \text{ Div } \overrightarrow{U} = 0 & \text{balance of mass} \\ \rho_0 \overrightarrow{U}_t + \overrightarrow{\text{Grad}}\, p = 0 & \text{balance of momentum} \\ p = c_0^2\, \rho & \text{equation of state} \end{cases}$$

where $\rho_0 = \rho_0(M)$, $c_0 = c_0(M)$, $M \in \mathbb{R}^3$ are given functions discontinuous on a dioptra, C^∞ outside the dioptra, and where $\rho = \rho(M,t)$, $\overrightarrow{U} = \overrightarrow{U}(M, t)$, $p = p(M, t)$ are the acoustic density, velocity vector and pressure respectively, in the case of a medium made of the juxtaposition of several immiscible fluids initially at rest. More generally we shall consider system (2) below that describes the case of a medium made of a juxtaposition of elastic solids (such a solid becomes a fluid by setting $\lambda_0 = c_0^2\, \rho_0$ and $\mu_0 = 0$) :

$$(2)\begin{cases} \overrightarrow{\xi}_t - \overrightarrow{U} = 0 & \text{definition} \\ \rho_t + \rho_0 \text{ Div } \overrightarrow{U} = 0 & \text{balance of mass} \\ \rho_0 \overrightarrow{U}_t - \text{Div } \overline{\Sigma} = 0 & \text{balance of momentum} \\ \overline{\Sigma} = \lambda_0 \text{ Div } \overrightarrow{\xi}\, \mathbf{1} + \mu_0\, (\overrightarrow{\text{Grad}\,\xi} + (\overrightarrow{\text{Grad}\,\xi}\,)^T) & \text{equation of state} \end{cases}$$

where $\overrightarrow{\xi} = \overrightarrow{\xi}(M,t)$ is the acoustic displacement vector and where $\overline{\Sigma} = \overline{\Sigma}(M,t)$ is the acoustic stress tensor ; $\lambda_0 = \lambda_0(M)$ and $\mu_0 = \mu_0(M)$ are the Lamé constants of the medium at rest.

Since the coefficients ρ_0, c_0, λ_0, μ_0 are discontinuous on a dioptra there appear (in the term $\rho_0 \text{ Div } \overrightarrow{U}$ in (1) ; in several other terms in (2)) products of distributions of the kind $Y . \delta$ (in 2 space dimensions one knows that $\overrightarrow{U}$ is indeed discontinuous on the dioptra). Our problem is to formulate the equations in $\mathcal{G}$, then to find existence - uniqueness (for given initial conditions) results and to compute the solutions.

6.1.2 <u>Statement of the equations</u>. How to state these equations in our setting ?, with = or $\approx$? From the general method exposed in §4.3 one states the basic conservation equations (and the first equation in (2)) with = in $\mathcal{G}$. Concerning the constitutive equations there is a basic difference with the case of shock waves considered up to now. There is no fast deformation of the medium inside the width of the dioptra : the interface evolves smoothly in the absence of acoustic perturbation (it is even at rest in the case of (1) and (2)) and then these acoustic perturbations are only linearly superposed to this evolution. This idea suggests to state also the constitutive equations with the (strong) equality in $\mathcal{G}$: the medium inside the infinitesimal width of the dioptra lies in the domain in which the classical constitutive equations are valid. This idea is supported by the facts that formal manipulations of these equations have never led to nonsense (Poirée [1, 2, 3, 4]) : as in 3.3.2 nonsense would be the indication that some equations should be stated with association. Does the statement of all equations in systems (1) (2) with equality in $\mathcal{G}$ guarantee existence and uniqueness of solutions ?

6.1.3 <u>An abstract existence - uniqueness result (1 dimensional case)</u>. Before the statement of the theorem we need a few technical definitions ; we limit ourselves to the $\mathcal{G}_s$ case to simplify the notation. For the reader who is not aware of chap 8 the notation $\mathcal{G}_s$ should be replaced by $\mathcal{G}$. A more precise statement of these results and detailed proofs can be found in Oberguggenberger [6] and Lafon-Oberguggenberger [1].

<u>Definition 1</u>. $U \in \mathcal{G}_s(\mathbb{R}^n)$ is called "locally bounded" iff it has a representative $u(\varepsilon,x)$ such that : for every bounded subset K of $\mathbb{R}^n$ there are $c > 0$ and $\eta > 0$ such that

$$(3) \qquad \sup_{x \in K} |u(\varepsilon,x)| \leq c \quad \text{if } 0 < \varepsilon < \eta.$$

U is called "globally bounded" when this bound holds when K is replaced by $\mathbb{R}^n$.

<u>Example.</u> Any classical $L^\infty_{loc}(\mathbb{R}^n)$ function is locally bounded when viewed as an element of $\mathcal{G}(\mathbb{R}^n)$ (def. 1 is extended at once to $\mathcal{G}(\mathbb{R}^n)$).

<u>Definition 2.</u> $U \in \mathcal{G}_s(\mathbb{R}^n)$ is called "locally of logarithmic growth" iff it has a representative $u(\varepsilon,x)$ such that : for every bounded subset K of $\mathbb{R}^n$ there are $c > 0$ and $\eta > 0$ such that

$$(4) \qquad \sup_{x \in K} |u(\varepsilon,x)| \leq c \, \text{Log} \frac{1}{\varepsilon} \text{ if } 0 < \varepsilon < \eta.$$

Example. Any element of $\mathcal{G}_s(\mathbb{R}^n)$ satisfies bounds of the type $\sup_{x \in K} |u(\varepsilon, x)| \leq \frac{c}{\varepsilon^N}$ for some $N \in \mathbb{N}$, if $0 < \varepsilon < \eta$. The bound (4) is more strict. However setting $\varepsilon' = \exp\left(-\frac{1}{\varepsilon^N}\right)$ and setting $v(\varepsilon', x) = u(\varepsilon, x)$ then v satisfies (4) on K. In usual cases the classes V, U $\in \mathcal{G}_s(\mathbb{R}^n)$ of v and u (respectively) are associated. In particular one proves easily (regularization by convolution with some function $\rho_{(\text{Log }(1/\varepsilon))^{-1}}, \rho \in \mathcal{D}, \int \rho = 1$) that for any distribution $T \in \mathcal{D}'(\mathbb{R}^n)$ there is $G \in \mathcal{G}(\mathbb{R}^n)$, locally of logarithmic growth, having the distribution T as macroscopic aspect. It seems that this logarithmic growth property is only a technical minor ingredient needed for the proof of the thm. below.

Theorem. Consider the linear (hyperbolic) system in one dimension

$$(5) \quad \begin{cases} (\partial_t + \Lambda(x,t)\partial_x) \, U(x,t) = F(x,t). \, U(x,t) + G(x,t) \\ U(x,0) = A(x) \end{cases}$$

where $x \in \mathbb{R}$, $t \in \mathbb{R}$, U and G are columns of n generalized functions of (x,t). We assume that Λ is diagonal and that all generalized functions concerned are real valued. We assume also that the coefficients of Λ are globally bounded and that the coefficients of $\partial_x \Lambda$ and F are locally of logarithmic growth. Then the Cauchy problem (5) has a unique solution in $\mathcal{G}_s(\mathbb{R}^2)$.

Remark. The proof and counterexamples (in case some assumptions are dropped) can be found in Oberguggenberger [6, 7, 8], see also Lafon [1]. The proof follows the classical method of integration along characteristics. It consists in considering the family of C^∞ problems depending on the parameter ε obtained by replacing Λ, F,G and A by representatives. Each of these problems has a C^∞ solution u_ε. Then one has to prove that the family $\{u_\varepsilon\}_\varepsilon$ lies in the reservoir of representatives of our generalized functions (the logarithmic growth assumption is used in this verification). Its class U is solution of (5). Uniqueness is obtained by considering the difference of two solutions and then by deriving suitable estimates for the difference. One can also prove that this result is coherent with classical results when they exist (L^1_{loc} case); also one can prove the well posedness - in the sense of $\mathcal{G}$ - of this problem, as this is proved in Biagioni [1] in the case of C^∞ coefficients.

6.1.4 Application to system(1) In one dimension system (1) stated with equalities in $\mathcal{G}$ reads as (c_0 and ρ_0 are strictly positive)

$$(1') \quad \begin{cases} (c_0^2 \, \rho_0)^{-1} p_t + u_x = 0 \\ \rho_0 \, u_t + p_x = 0 \end{cases}$$

with u and p given at $x = x_0 < 0$. Diagonalization is done by introducing new dependent variables $(v_1, v_2) \in \mathcal{G}(\mathbb{R}^2)$ by

$$\begin{cases} u = v_1 - v_2 \\ p = c_0\, \rho_0(v_1 + v_2) \end{cases}$$

and one obtains the system

$$(1'') \quad \partial_t \begin{pmatrix} v_1 \\ v_2 \end{pmatrix} + \begin{pmatrix} c_0 & 0 \\ 0 & c_0 \end{pmatrix} \partial_x \begin{pmatrix} v_1 \\ v_2 \end{pmatrix} = c_0\, \mu \begin{pmatrix} 1 & 1 \\ 1 & 1 \end{pmatrix} \begin{pmatrix} v_1 \\ v_2 \end{pmatrix}$$

where $\mu = -\frac{1}{2}(c_0\, \rho_0)^{-1} \partial_x(c_0\, \rho_0)$. The above existence - uniqueness result applies to $(1'')$.

6.1.5 <u>Problems on the macroscopic aspects ot the solutions.</u> However, there remains a basic problem : from the physical (macroscopic) viewpoint the coefficients $\rho_0(x)$ and $c_0(x)$ are known as classical functions discontinuous on the dioptra. As the reader knows now very well there are several ways in $\mathcal{G}_s$ to interpret classical functions as generalized functions (particularly at their points of discontinuity - let us repeat that the canonical inclusion $\mathcal{D}' \subset \mathcal{G}$ has to be left here since it is not physically justified and since it may also be destroyed by the requirement of logarithmic growth). So the above uniqueness result leaves open the following question. Let $(c_{0,1}, \rho_{0,1})$ and $(c_{0,2}, \rho_{0,2})$ be two pairs of elements of $\mathcal{G}_s(\mathbb{R})$ locally bounded and of local logarithmic growth) associated with a given pair (c_0, ρ_0) of classical functions (discontinuous on the dioptra, C^∞ outside it). For the same initial condition in $\mathcal{G}$ - or more generally for associated initial conditions - let (ρ_1, u_1, p_1) and (ρ_2, u_2, p_2) be the corresponding solutions (obtained from the above theorem). Are (ρ_1, u_1, p_1) and $(\rho_2, u_2, p_2) \in \mathcal{G}(\mathbb{R}^2)$ associated ? Another basic problem is : is there a choice of the coefficients and of the initial conditions (among the various elements of $\mathcal{G}_s$ having a given macroscopic aspect, and satisfying the assumptions of the abstract existence uniqueness theorem, such that the corresponding solution has the macroscopic aspect expected from physics ? (the general theorem gives no piece of information on the solution). In the case of $(1')$ these questions are solved as follows.

6.1.6. <u>The classical aspect of the solution : continuity of u and p in one dimension.</u> Let us assume that the discontinuity of the medium lies at $x = 0$. First we begin with a classical calculation from physics: let us seek $W = \rho, u, p$ of the form

$$(6) \quad W(x,t) = W_c(x,t) + [W](t)\, H_w(x,t)$$

where H_w is an Heaviside generalized function, where W_c is the continuous part of W and where $[W]$ is the jump of W at $x = 0$. From physical reasons we assume

(6') $\begin{cases} W_c \text{ continuous and bounded, } (W_c)_t \text{ and } (W_c)_x \text{ bounded} \\ \qquad\qquad\qquad\qquad\qquad\qquad \text{on both sides of the discontinuity} \\ [W] \text{ and } [W]_t \text{ bounded} \\ H_w \text{ bounded} \end{cases}$.

Putting (6) into (1') and using (6') one obtains easily (Colombeau [11]) that [u] and [p] are associated with 0 : u and p have the macroscopic aspect of continuous functions, which is well known in this case (Poirée [1, 2, 3, 4]). In fact this result (without the use of the extra conditions (6')) follows from a much deeper study of the abstract theorem 6.1.3 in the particular case of (1'). That is what we describe now.

6.1.7 The classical aspect of the solutions of (1)

The classical piecewise C^∞ functions ρ_0 and c_0^2 are obtained from experiments or observations. Any generalized function $\tilde{\rho}_0$ (or $\tilde{c}_0^2$) associated with ρ_0 (and c_0^2 respectively) contains more physical information than ρ_0 (and c_0^2 respectively). Indeed in our more precise setting the physical situation has to be described a priori by generalized functions whose macroscopic aspects are the classical functions mentioned above. But the "microscopic" observation of the jump of $\tilde{\rho}_0$ and $\tilde{c}_0^2$ has not been done, and so we do not know the "profile" of $\tilde{\rho}_0$ and $\tilde{c}_0^2$ on their discontinuity.

More refined physical pieces of information would be requested to know this profile. Fortunately we are going to show that such a knowledge is not needed in the particular case considered here. Indeed we have a rather large choice of generalized functions $\tilde{\rho}_0$ and $\tilde{c}_0^2$ which are associated with ρ_0 and c_0^2, and could as well represent the phenomenon considered here. Does this choice influence the solution ? An answer "yes" would mean that indeed we have not enough physical pieces of information to predict the solution. Fortunately, the answer will be "no", modulo the association.

Proposition 1. *Let ρ_0 and c_0^2 be given classical (discontinuous) functions ; let $\tilde{\rho}_0$ and $\tilde{c}_0^2$ be generalized functions associated with ρ_0 and c_0^2 respectively. Let us make on $\tilde{\rho}_0$ and $\tilde{c}_0^2$ the reasonable assumption that their representatives are bounded in absolute value and from below by a strictly positive constant depending on each bounded subset of $\mathbb{R}$, independently of ε. Let us assume the boundary data are $\mathcal{C}^1$. Then the various solutions (u,p) depending on $\tilde{\rho}_0$ and $\tilde{c}_0^2$ are associated with each other. In other words, modulo the association, the solutions of (1') depend only on ρ_0 and c_0^2.*

Remark . In the case of other systems of equations (that are not physical models) one finds that the macroscopic aspect of the solutions in $\mathcal{G}$ (obtained in general form the above theorem) depends on microscopic properties of the coefficients, see Oberguggenberger [6, 7].

Proposition 2. *Under the conditions of prop. 1 above and if the boundary data are $\mathcal{C}^\infty$, the solutions of (1') are associated with a pair $(\overline{u},\overline{p})$ of piecewise C^∞ functions which are both continuous at* $x = x_0$.

One recovers the solution expected by physicists. The proofs of prop. 1 and 2 are based on arguments using the H^1_{loc} theory, see Oberguggenberger [6, 7, 8].

- Remark. When one knows this last result (continuity of u and p at $x = x_0$) then one can solve (existence and uniqueness) system (1') in the classical setting of the Sobolev spaces H^1_{loc} $(\mathbb{R}^2)$. However this is only a very special remark : this last setting cannot explain "formal" calculations done by physicists to manipulate the system of acoustics : derivation of propagation equations, change of representations (lagrangian, eulerian) and further it is based on assumptions (on the solutions) which are not necessarily realistic, or are not obvious, in more complicated cases, see (11') in chap. 2.

6.1.8 - The multi - dimensional case

Although considerably more technical - and more interesting since it is predictive of new physical results, - it follows essentially the same pattern. A general existence and uniqueness theorem can be proved, generalizing thm. 6.1.3. Already in dimension two the interface may be a rather complicated curve, with singular points. The method of proof still relies upon the resolution of a family of C^∞ problems depending on the parameter ε, and on suitable uniform bounds in ε, see Lafon - Oberguggenberger [1].

As in the one dimensional case, this implies existence (and uniqueness when initial data are prescribed) of solutions in $\mathcal{G}(\mathbb{R}^2)$ for the system

$$(7)\begin{cases} p_t + \rho_0 \, \overrightarrow{Div u} = 0 \\ \rho_0 \, \overrightarrow{u}_t + \overrightarrow{Grad} \, p = 0 \\ p = c_0^2 \, \rho \end{cases}$$

generalizing system (4'), see Lafon - Oberguggenberger [1]. This also applies to the elastic solid case in two space dimensions, see Laurens [1].

6.1.9 - Problems

"Concrete results" such as those in 6.1.7, should be developed for the rather large number of systems of equations of linear and weakly nonlinear acoustics, as well as, in dimension ≥ 2, for various possible kinds of more or less irregular interfaces. From various numerical tests and from Laurens [2] we expect that-under natural assumptions on ρ_0 and c_0^2 - the "abstract solution" has a classical macroscopic aspect as in 6.1.7 for the one dimensional case. Note also that this kind of problem is not at all limited to acoustics ; in fact it concerns various kinds of linear wave propagation in a medium with discontinuous characteristics, see for instance Lafon [1] for an electron transport problem.

6.1.10 An interesting mathematical example of a single linear equation with a discontinuous coefficient : the Hurd-Sattinger equation, see Oberguggenberger [7].

It is the equation

$$(8) \quad \begin{cases} \partial_t v(x,t) - \partial_x (H(x) v(x,t)) = 0 \\ v(x,0) = 1 \end{cases}$$

with H "the" (in the classical sense) Heaviside function (or in $\mathcal{G}$ "an" Heaviside function). If v is a solution : for $x < 0$ we have $\partial_t v = 0$ i.e. $v \equiv 1$, for $x > 0$ we have $(\partial_t - \partial_x) v = 0$ which implies easily $v \equiv 1$; thus $v(x,t) = 1$ if $x \neq 0$. Then if v is a classical function $\partial_t v = \partial_x H(x)$ thus $\partial_t v = \delta(x)$ and so we find $v(x,t) = 1 + t \delta(x)$, $t > 0$ which is not a classical function. Oberguggenberger [7] has proved that (8) stated with equality in $\mathcal{G}$ has a unique solution in $\mathcal{G}$ ($\mathbb{R} \times [o, +\infty]$) and that further this solution has a macroscopic aspect which is $1 + t\delta(x)$. Thus (8) involves really a product of distributions of the kind Y. δ. In Oberguggenberger [7] one finds also many other mathematical examples of linear equations with singular coefficients.

6.1.11 Problem Consider the equation

$$(8') \quad \begin{cases} \partial_t v(x,t) - \partial_x (H(x) v(x,t)) = 0 \\ v(x,o) = A(x) \end{cases}$$

Assume that $A \in \mathcal{G}(\mathbb{R})$ has as macroscopic aspect a function $a \in L^1_{loc}(\mathbb{R})$. Then prove that the solution of (8') admits an associated distribution given by the formula

$$(8") \quad Y(-x)\, a(x) + Y(x)\, Y(x+t)\, a(x+t) + \delta(x)\, Y(t) \int_0^t a(\xi)\, d\xi$$

where Y and δ denote respectively the Heaviside and Dirac distributions (the proof is in Oberguggenberger [7]). From formula (9) the macroscopic aspect of the solution can be represented as follows :

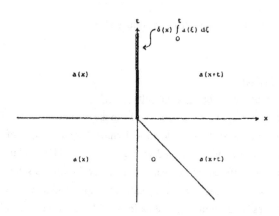

<u>6.1.12 Example.</u> The solution in 6.1.11 has a macroscopic aspect independent on H and A (that have the macroscopic aspects of the distributions Y and a respectively). Here is an example extracted from Baraille et al [1] in which the solution has a macroscopic aspect that depends on the microscopic properties of the data. Consider the equation

$$\rho u_t + p_x = 0$$

where $\rho(x,t) = \Delta\rho\, H(x) + \rho_g$, given, $\rho > 0$, $\rho \in \mathcal{G}(\mathbb{R})$,

$p(x,t) = \Delta p\, K(x) + p_g$ given,

$u(x,o) = \Delta u\, L(x) + u_g$ given,

in which H, K, L are given Heaviside generalized functions. Seek a solution

$u(x,t) = \Delta u\, L(x) + u_g + \alpha t\delta(x)$

with δ a Dirac generalized function. Putting these expressions in the equations we get the relation

$$\alpha\delta(x) = \frac{-\Delta p\, K'(x)}{\Delta\rho\, H(x) + \rho_g}$$

and, setting $H\delta \approx A\delta$, we get the relation $\alpha = \dfrac{-\Delta p}{A\Delta\rho + \rho_g}$. The macroscopic aspect of the solution u

relies on the number α, which depends on A. The relation above implies

$$\alpha(H\delta)\,(x) = \frac{-\Delta p\ H(x)\ K'(x)}{\Delta\rho\ H(x) + \rho_g}$$

and thus

$$\alpha\,A = -\Delta p \int_{-\infty}^{+\infty} \frac{H(x)\ K'(x)}{\Delta p\ H(x) + \rho_g}\,dx.$$

Setting $H = K^n$ one checks that this leads to different values of α.

§6.2 A NUMERICAL METHOD

6.2.1 A numerical adaptation of the mathematical results in §6.1

Mere adaptations of classical numerical methods from the continuous coefficients case amount intuitively to obtaining solutions of system (1) stated with three associations, or of systems obtained from (1) by formal manipulations, and also stated with associations. This explains the observation (next section) that one obtains different numerical results according to the scheme in use, usually different from the correct solution (in cases this last solution is known). One recovers numerically the fact that this system involves ambiguous multiplications of distributions. In order to discover a numerical method giving a solution of the system stated with three (strong) equalities, one has to consider the ideas corresponding to the concept of (strong) equality. This concept expresses that this kind of equality is still valid in a scale smaller than the width of the dioptra : thus the numerical scheme has to register the fact that the equations are valid even inside the width of the dioptra.

The classical methods from the continuous coefficients case use a "brutal" discretization of the dioptra (fig. 1).

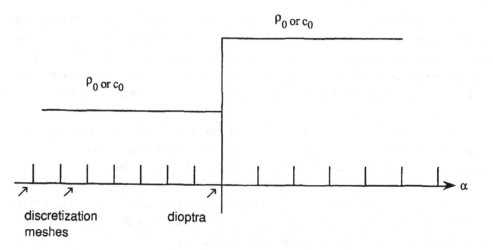

ρ_0 or c_0

ρ_0 or c_0

α

discretization
meshes

dioptra

Figure 1 : *Brutal discretization of the dioptra in one space dimension*

Then since in these methods the dioptra has no width, they do not register the valididy of the equations inside its width. They only register the validity of the equations on both sides. Intuitively this corresponds to our concept of association.

In order to discretize the equations with (strong) equality, one has to introduce numerically the concept of width of the dioptra. One can represent this width by one mesh, let ρ_0 and c_0 be continuous (by a continuous junction) in this mesh and divide it into a sufficient number of "small meshes" so as to discretize the equations inside this width, fig.2.

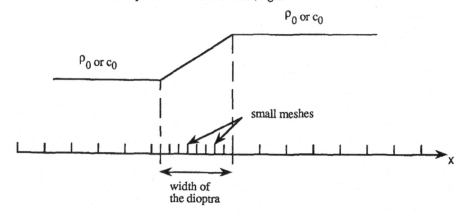

ρ_0 or c_0

ρ_0 or c_0

small meshes

x

width of
the dioptra

Figure 2 : *A discretization that permits to register the validity of the equations inside the width of the dioptra.*

A few numerical results are described in the next section. More generally one has observed that all numerical schemes, when they are treated by this method of small meshes, give the same solution, which is the solution of the system stated with three strong equalities. This result does not depend on the discretization in use as starting point : the small mesh method is the only cause of this good result. Thus these numerical results bring exactly a confirmation of our analysis in §6.1.

In order to find a numerical scheme that would be both correct and not too much expensive in computation time and memory, one uses the obviously natural following technique. At first one discretizes the width of the dioptra with small meshes (fig. 3 in the case of a diedra). Then :

(1) one arbitrarily chooses a natural scheme of the continuous coefficients case and one treats it by the small mesh technique till the solution obtained is not any longer modified (up to the precision under consideration) by an increase in the number of small meshes. Then one has obtained the desired solution (the sequel is only devoted to improvement of the calculation cost).

(2) one tests various natural schemes of the continuous coefficients case, without small meshes : they give different results. One selects the scheme which gives the result closer to the correct solution obtained in (1) above. Then one treats this scheme by the small meshes method by taking only a minimum number of small meshes. The scheme so selected is in general much more efficient than the one chosen at random in (1) : indeed the number of small meshes needed to obtain the correct solution, and so the cost of calculations, depends on the discretization scheme (although the final result, provided one uses a sufficient number of small meshes, does not depend on it).

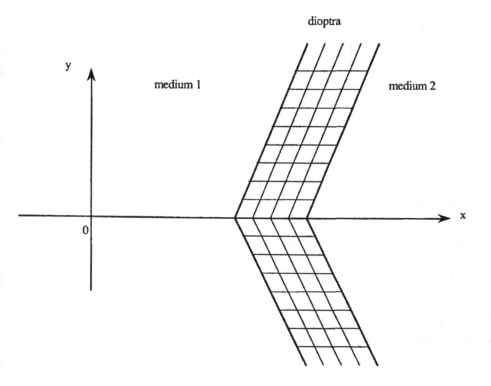

Figure 3 : Discretization of a diedra by small meshes

6.2.2 Numerical tests in the one dimensional case. We test two arbitrary discretizations for system (1'). The first one is an explicit scheme : knowing

$\{u_n^i\}_{i\in\mathbb{Z}}$ and $\{p_i^n\}_{i\in\mathbb{Z}}$ one computes u_i^{n+1} and p_i^{n+1} from the formulas :

$$(9)\begin{cases} m(p_i^n) = \frac{1}{4}(p_{i-1}^n + 2p_i^n + p_{i+1}^n) \\[2mm] m(u_i^n) = \frac{1}{4}(u_{i-1}^n + 2u_i^n + u_{i+1}^n) \\[2mm] p_i^{n+1} = m(p_i^n) - \frac{r}{2}\rho_o^i (c_o^i)^2 (u_{i+1}^n - u_{i-1}^n) \\[2mm] u_i^{n+1} = m(u_i^n) - \frac{r}{2\rho_o^i} (p_{i+1}^n - p_{i-1}^n). \end{cases}$$

We also consider an implicit scheme (retrograd techniques). Stating system (13) under the form.

$$(9') \quad \begin{pmatrix} p \\ u \end{pmatrix}_t + \begin{pmatrix} 0 & \rho_0 c_0^2 \\ \dfrac{1}{\rho_0} & 0 \end{pmatrix} \begin{pmatrix} p \\ u \end{pmatrix}_x = 0$$

and doing the change of unknown variables

$$\begin{cases} R = p - \rho_0 c_0 u \\ Q = p + \rho_0 c_0 u \end{cases}$$

one obtains the system

$$(9'') \quad \begin{cases} R_t - c_0 R_x = \dfrac{1}{2\rho_0} (\rho_0 c_0)_x \, (Q-R) \\[2mm] Q_t + c_0 Q_x = \dfrac{1}{2\rho_0} (\rho_0 c_0)_x \, (Q-R). \end{cases}$$

This system is treated by a classical implicit propagation - convection technique (Barka [1]). After some calculations one obtains the formulas

$$(10) \begin{cases} P_\eta(R_i^n) = \dfrac{(c_0)_{i+1}^n \Delta t R_{i+1}^n + [h(c_0)_i^n \Delta t] R_i^n}{h + [(c_0)_{i+1}^n - (c_0)_i^n] \Delta t} \\[4mm] P_\xi(Q_i^n) = \dfrac{(c_0)_{i-1}^n \Delta t \, Q_{i-1}^n + [h - (c_0)_i^n \Delta t] Q_i^n}{h + [(c_0)_{i-1}^n - (c_0)_i^n] \, \Delta t} \\[4mm] R_i^{n+1} = (1 - r\gamma_i) P_\eta \, (R_i^n) + r \, \gamma_i \, P_\xi \, (Q_i^n) \\[2mm] Q_i^{n+1} = - r\gamma_i \, P_\eta \, (R_i^n) + (1 + r \, \gamma_i \,) \, P_\xi \, (Q_i^n) \\[2mm] \gamma_i = \dfrac{(\alpha_i + \beta_i)}{2} \\[3mm] \alpha_i = \dfrac{(\rho_0 c_0)_{i+1} - (\rho_0 c_0)_i}{2(\rho_0)_{i+1}}, \; \beta_i = \dfrac{(\rho_0 c_0)_i - (\rho_0 c_0)_{i-1}}{2(\rho_0)_i}. \end{cases}$$

For the numerical results in fig.8,9,10 one considers a square wave proceeding towards the dioptra (fig. 4) :

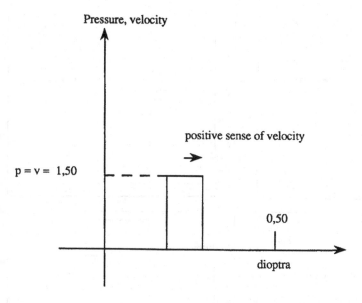

Figure 4 : *Initial data (square wave) in velocity and pressure*

In the figure 5,6,7 one chooses $\rho_0 = c_0 = 1$ if $x < 0,5$ and $\rho_0 = c_0 = 1,5$ if $x > 0,5$.

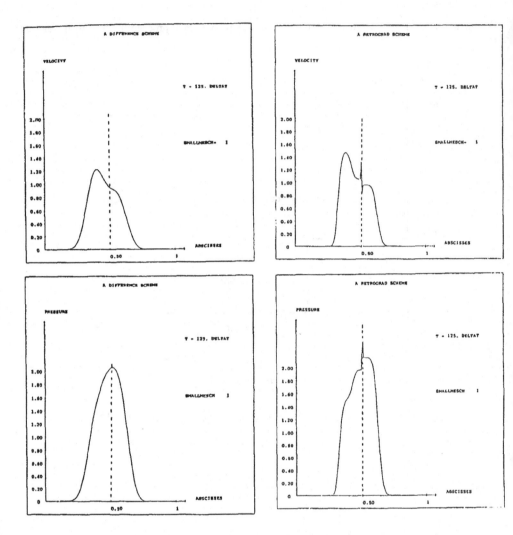

Figure 5 : _The two schemes give neatly different results when one does not use the small meshes method_

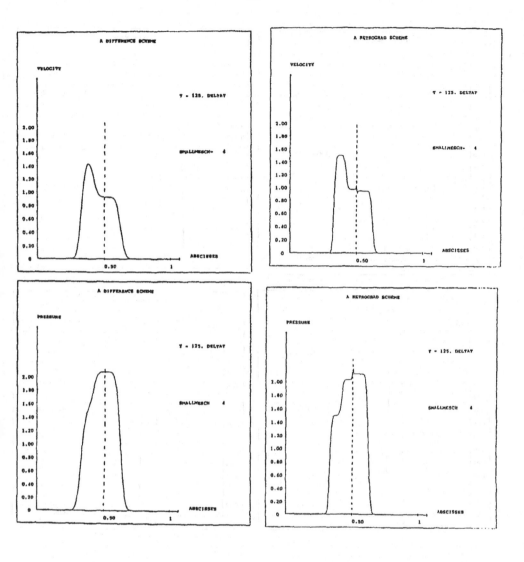

<u>Figure 6</u> : *Results obtained with 4 small meshes*

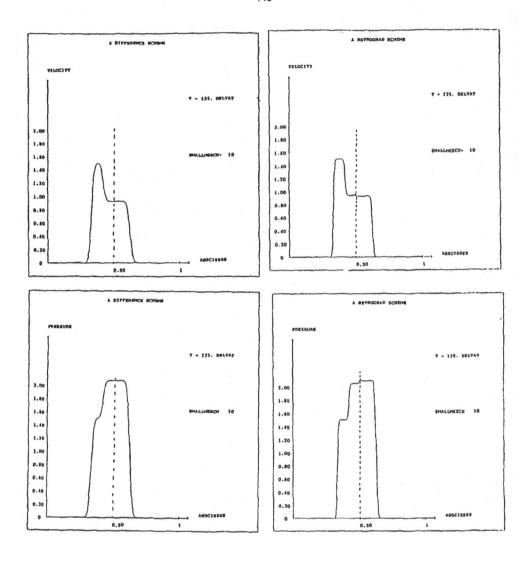

Figure 7 : *Results obtained with 10 small meshes. The evolution that can be observed in fig. 8. 9. 10 shows that the increase in the number of small meshes produces, at the limit, the same solution. One has observed continuation in this evolution when one still increases the number of small meshes, see Barka [1], Barka-Colombeau-Perrot [1].*

6.2.3 <u>Numerical schemes in the case of a two dimensional diedra.</u> The method can easily be adapted to the 2 dimensional diedra. An obvious way is to have only one kind of meshes but to state that the jump of ρ_0, c_0 takes place on 5, or 10, or more meshes (width of the dioptra) ; the programming is immediate (this is indeed what has been done in fig 5,6,7) but this method is expensive in computation time and memory space. Therefore one prefers to consider that the jump of ρ_0 and c_0 takes place on one mesh, that one divides into 5, 10,... small meshes (in the sense parallel to the bissectrix of the diedra, see fig. 3). One has to be cautious in the various interpolations needed at the interface between the usual (large meshes) and the small meshes. Various more or less accurate schemes have been developed both for the fluid/fluid system 1 and for the fluid/ solid system (2), see Barka [1], Barka - Colombeau - Perrot [1], Laurens [1], Colombeau - Laurens - Perrot [1] where numerical results are reported.

Then, if one admits that the above "small mesh method" gives the correct solution, on can test numerically more efficient (without small meshes) methods, and thus find other methods that give the correct solution. In this process the small mesh method plays the role of a test for other methods. However it seems that one can avoid the problem of multiplication of distributions in linear acoustics (Laurens [2]). Thus the above methods could be more interesting in more evolved cases in which this problem cannot be avoided. Here are a few numerical results obtained from this method as numerical solutions of system (16") of chap 2. (L.S. Chadli, still unpublished). An elastic solid diedra is immerger into water, and a thin acoustic signal is sent on the edge V of the diedra. One observes one or two reflected acoustic pencils P_1, P_2 (one for plexiglass, two for aluminium, see below).

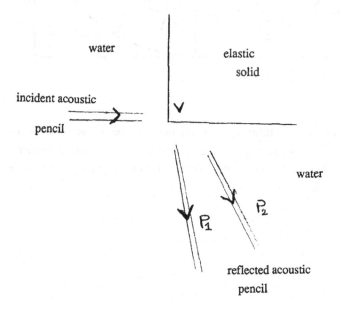

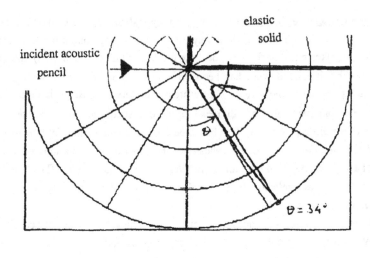

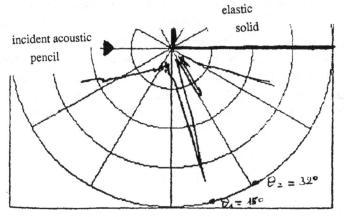

Figure 8. Experimental results (De Billy [1] describing measurements of the maximum of absolute value of acoustic pressure according to the angle θ. On top the solid is made of plexiglass and one observes only one pencil at $\theta = 34°$. Below the solid is made of aluminium and one observes two pencils at $\theta_1 = 15°$ and $\theta_2 = 32°$.

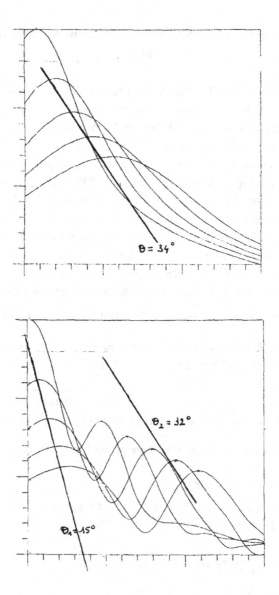

Figure 9. Numerical results : the maximum of absolute value of acoustic pressure has been registered on 5 straight lines parallel to the horizontal side of the diedra, which gives at once the reflected pencils and their angles. On top case of plexiglass and below case of aluminium.

Chapter 7. The canonical Hamiltonian formalism of interacting quantum fields.

§7.1. <u>GENERALITIES AND THE FOCK SPACE</u>. The aim of this chapter is to show - on the example of a celebrated theory of physics - the kind of "formal calculations on functions" used by physicists. The calculations presented in this chapter go back to the end of the twenties, when Dirac [2] founded Quantum Field Theory. This was done by quantization of the classical electromagnetic field, i.e. the replacement of the classical field functions by operator valued functions ; these operators act on a Hilbert space, the space of states (called Fock space). A state represents a certain (finite or infinite) number of particles at a given time. Quantum Electrodynamics deals with three kinds of particles : the photon (particle of light, massless and without electric charge), the electron (with a certain mass and an electric charge - e) and its antiparticle the positron (same mass, but electric charge + e), and describes their mutual transformations when they collide, see Kastler [1], Bogoliubov and Shirkov [1]. For simplification we consider here a model in which only one kind of particle is involved ; these particles (neutral scalar bosons with mass m > 0) are supposed to interact with themselves according to the simplest possible nonlinear equation $\left(\frac{\partial^2}{\partial t^2} - \Delta + m^2\right) u + gu^3 = 0$). This model presents all main mathematical features of Quantum Field Theory, concerning manipulations on "generalized functions". The scenery takes place in a specific Hilbert space, the Fock space, that we describe now.

Let $L^2(\mathbb{R}^{3n})$ denote the Hilbert space of all square integrable complex valued functions on $\mathbb{R}^{3n}$ (with respect to the Lebesgue measure). We denote by $L_s^2((\mathbb{R}^3)^n)$ the closed subspace of $L^2(\mathbb{R}^{3n})$ of those functions which are (almost everywhere) symmetric functions of their n arguments in $\mathbb{R}^3$, i. e.

$$f(x_{\sigma_1}, ..., x_{\sigma_n}) = f(x_1, ..., x_n)$$

for almost all $(x_1, ..., x_n)$ and for all permutations σ of the set $\{1, ..., n\}$. By definition the Fock space $\mathbb{F}$ is the (infinite) Hilbertian direct sum

$$\mathbb{F} = \mathbb{C} \oplus L^2(\mathbb{R}^3) \oplus L_s^2((\mathbb{R}^3)^2) \oplus ... \oplus L_s^2((\mathbb{R}^3)^n) \oplus ...$$

Let us recall that the Hilbertian direct sum $S = \bigoplus_{i=1}^{\infty} H_i$, H_i Hilbert spaces, is the family of all infinite sequences (x_i), $x_i \in H_i$ $\forall i$, such that $\sum_{i=1}^{\infty} \|x_i\|_{H_i}^2 < +\infty$ where $\|x_i\|_{H_i}$ is the norm of x_i in H_i. S is a Hilbert space with the inner product $<(x_i),(y_i)>= \sum_{i=1}^{\infty} <x_i, y_i>_{H_i}$ if $<,>_{H_i}$ denotes the inner product of H_i. Any element K of $\mathbb{F}$ may be represented by the infinite column

$$K = \begin{pmatrix} K_0 \\ K_1(x) \\ K_2(x_1,x_2) \\ \cdot \\ \cdot \\ \cdot \\ K_n(x_1,...,x_n). \\ \cdot \\ \cdot \end{pmatrix}$$

where $K_0 \in \mathbb{C}$ and where $K_n(x_1,...,x_n)$ stands for the function

$$x_1, ..., x_n \rightarrow K_n(x_1, ..., x_n)$$

(and is such that $K_n(x_1,...,x_n) = K_n(x_{\sigma_1},...,x_{\sigma_n})$ for all permutations σ of the set $\{1,...,n\}$). Then

$$\|K\|_{\mathbb{F}}^2 = |K_0|^2 + \sum_{n=1}^{+\infty} \int |K_n(x_1,...,x_n)|^2 \, dx_1...dx_n < +\infty.$$

A state K with only one nonzero element, i. e. $K_n = 0$ if $n \neq p$ and $K_p \neq 0$ is called a state with p

particles exactly. $K_o = \begin{pmatrix} 1 \\ 0 \\ 0 \\ \cdot \\ 0 \\ \cdot \end{pmatrix}$ is called the vacuum state (state with 0 particle exactly).

The states that are physically interpretable are those K of norm 1 in $\mathbb{F}$. Then $\|K_n\|_{L^2(\mathbb{R}^{3n})}^2$ denotes the

probability that this state contains n particles exactly ; in general a state contains an unspecified number of particles. Basic concepts accompanying the Fock space are the creation and annihilation operators $a^+(\phi)$ and $a^-(\phi)$ respectively, $\phi \in L^2(\mathbb{R}^3)$ variable. $a^+(\phi)$ is defined by

$$(1) \quad a^+(\phi) \begin{pmatrix} K_0 \\ K_1(x) \\ \cdot \\ \cdot \\ \cdot \\ K_n(x_1,...,x_n) \\ \cdot \\ \cdot \end{pmatrix} = \begin{pmatrix} 0 \\ K_0\phi(x) \\ \sqrt{2} \, \text{Sym} \, \phi(x_1) K_1(x_2) \\ \cdot \\ \cdot \\ \cdot \\ \sqrt{n} \, \text{Sym} \, \phi(x_1) K_{n-1}(x_2,...,x_n) \\ \cdot \\ \cdot \end{pmatrix}.$$

where Sym. is the operator of symmetrization of a function $(\text{Sym } f\ (x_1,...,x_n) = \frac{1}{n!}$

$\sum_{\sigma \in G_n} f(x_{\sigma_1},...,x_{\sigma_n})$ if G_n is the set of the n! permutations of the set $\{1,...,n\}$). Therefore a^+ (ϕ)

transforms a state with p particles exactly into a state with (p + 1) particles exactly ; thus its name of creation operator (of a particule in the state $\phi \in L^2(\mathbb{R}^3)$). The annihilation operator a^- (ϕ), $\phi \in L^2(\mathbb{R}^3)$ variable, is defined by

$$
(2) \quad a^-\ (\phi) \begin{pmatrix} K_o \\ K_1(x) \\ \cdot \\ \cdot \\ K_n(x_1,...,x_n) \\ \cdot \\ \cdot \end{pmatrix} = \begin{pmatrix} \int K_1(\xi)\ \phi(\xi)\ d\xi \\ \sqrt{2}\ \int K_2(x,\xi)\ \phi(\xi)\ d\xi \\ \cdot \\ \cdot \\ \sqrt{n+1}\ \int K_{n+1}(x_1,...,x_n,\xi)\phi(\xi)\ d\xi \\ \cdot \end{pmatrix}
$$

$a^-(\phi)$ transforms a state with p particles exactly into a state with p - 1 particles exactly, thus its name of annihilation operator. These operators are unbounded operators in the Fock space $\mathbb{F}$ and their domain contains the states with a finite number of particles (i. e. the states such that $K_n = 0$ for large enough n). The brackets [A, B] of two operators A and B on $\mathbb{F}$ is defined by

$$[A, B] = A \circ B - B \circ A$$

(where o denotes the composition of operators). From (1) and (2) one obtains easily the "canonical commutation relations" valid for any $\phi_1, \phi_2 \in L^2(\mathbb{R}^3)$:

$$
(3) \begin{cases} [a^+(\phi_1), a^+(\phi_2)] = 0 = [a^-(\phi_1), a^-(\phi_2)] \\ [a^-(\phi_1),\ a^+(\phi_2)] = \int \phi_1(x)\phi_2(x)\ dx\ \text{Id} \end{cases}
$$

where Id denotes the identity operator on $\mathbb{F}$. If $k \in \mathbb{R}^3$, the creation and annihilation operators a^+ (k) and a^- (k) are obtained from a^+ (ϕ) and a^- (ϕ) respectively by replacing the function ϕ by the function $x \to \delta(x - k)$ where δ is the Dirac delta function (and thus $a^\pm(\phi) = \int a^\pm(k)\ \phi(k)\ dk$). Of course $a^\pm(k)$ make sense by considering that $a^\pm(k)$ are not usual functions, but distributions in the variable $k \in \mathbb{R}^3$. Then (3) becomes, for any $k_1, k_2 \in \mathbb{R}^3$,

$$(3) \begin{cases} [a^+(k_1), a^+(k_2)] = 0 = [a^-(k_1), a^-(k_2)] \\ [a^-(k_1), a^+(k_2)] = \delta(k_1 - k_2)\mathrm{Id}. \end{cases}$$

§7. 2 THE FREE FIELD. The free field operator is given by the formula

$$(4) \quad A_0(x,t) = (2\pi)^{\frac{-3}{2}} 2^{\frac{-1}{2}} \int_{k \in \mathbb{R}^3} (k^\circ)^{\frac{-1}{2}} \{e^{ik^\circ t} e^{-ikx} a^+(k) + e^{-ik^\circ t} e^{ikx} a^-(k)\} \, dk$$

where $k, x \in \mathbb{R}^3$, $k^\circ = (k^2 + m^2)^{\frac{1}{2}}$, $k^2 = (k_1)^2 + (k_2)^2 + (k_3)^2$ if $k = (k_1, k_2, k_3) \in \mathbb{R}^3$;
It has been proved (Whigtman [1], Jost [1]) that A_0 is not an (operator valued) function in any "reasonable" sense.

Let $\psi \in \mathcal{D}(\mathbb{R}^3)$ be given. We set

$$(5) \quad \mathcal{F} \psi(k) = (2\pi)^{\frac{-3}{2}} \int_{\lambda \in \mathbb{R}^3} e^{-ik\lambda} \psi(\lambda) d\lambda$$

(this is the definition of the Fourier transform used by physicists ; it differs slightly from the definition used by mathematicians, which is $\mathcal{F} \psi(k) = \int e^{-2i\pi k\lambda} \psi(\lambda) \, d\lambda$). In this chapter we shall use the definition of physicists. Setting

$$(6) \qquad A_0(\psi, t) = \int A_0(x, t) \, \psi(x) \, dx$$

a "formal" use of Fubini's theorem gives at once

$$(7) \qquad A_0(\psi, t) = 2^{-\frac{1}{2}} \{a^+[k \to (k^\circ)^{-\frac{1}{2}} e^{ik^\circ t} \mathcal{F} \psi(k)] + a^-[k \to (k^\circ)^{-\frac{1}{2}} e^{-ik^\circ t} \mathcal{F} \psi(-k)]\}.$$

Since $a^\pm(\phi)$ are (unbounded) operators on $\mathbb{F}$ (with dense domain) if $\phi \in L^2(\mathbb{R}^3)$ it follows at once that for all $\psi \in \mathcal{D}(\mathbb{R}^3)$ then $A_0(\psi, t)$ defined by (7) is an operator on $\mathbb{F}$ (unbounded, with dense domain). Therefore, for each fixed t, $A_0(x, t)$ is in fact a distribution in the x – variable, and not a function. It is immediate to check that (7) is a solution, in the sense of distributions, of the linear wave equation

$$(8) \qquad \left(-\frac{\partial^2}{\partial t^2} + \sum_{\mu=1}^{3} \frac{\partial^2}{\partial x_\mu^2} - m^2 \right) A_0(x,t) = 0.$$

In fact this verification follows at once from the factors $e^{ik^{\circ}t}\, e^{-ikx}$ and $e^{-ik^{\circ}t}\, e^{ikx}$ in (4). One can immediately find a dense domain D in $\mathbb{F}$ such that $A_0(\psi, t)$ and $\frac{\partial}{\partial t} A_0(\psi, t)$ map D into D for any $\psi \in$ $\mathfrak{D}(\mathbb{R}^3)$ (it suffices to choose for D the set of all states with a finite number of particles). From (3', 4) or (3, 7) one gets at once the commutation relations of the free fields, where we set $\pi_0(x, t) = \frac{\partial}{\partial t}$ $A_0(x, t)$:

(9) $\begin{cases} [A_0(x, t), A_0(x', t)] = 0 = [\pi_0(x, t), \pi_0(x', t)] \\ [A_0(x, t), \pi_0(x', t)] = i\, \delta(x - x')\ \text{Id.} \end{cases}$

Remark. In classical mechanics and calculus of variation the Euler equations are obtained by minimization of functionals ; if $\mathcal{L}(x, t)$ is the Lagrangian density then the minimization of the functional $\int_{\mathbb{R}_4} \mathcal{L}(x, t)\, dx\, dt$ leads to the Euler equation (where $x = (x_1, x_2, x_3)$, $x_0 = t$)

(10)
$$\sum_{\mu=0}^{3} \frac{\partial}{\partial x_\mu} \frac{\partial \mathcal{L}}{\partial(\frac{\partial A}{\partial x_\mu})} - \frac{\partial \mathcal{L}}{\partial A} = 0.$$

In our case setting

(11) $\mathcal{L}(x,t) = -\frac{1}{2}\left\{ \sum_{\mu=1}^{3} \left(\frac{\partial A}{\partial x_\mu}\right)^2 - \left(\frac{\partial A}{\partial t}\right)^2 + m^2\, A^2 \right\}(x,t)$

where A is some –a priori - unknown "function" of (x,t)(whose value for each (x,t) is assumed to be an operator on $\mathbb{F}$), then (10) and (11) give the equation (8) for A. Since the solution A_0 is not a function, but a distribution, one can check that formula (11) is meaningless within distribution theory. This is not a severe drawback since one can content oneself with the explicit formula for A_0, and drop the Lagrangian density as useless.

Remark. The energy operator P_0 is defined by the formula

(12) $\quad P_0 = \int_{k \in \mathbb{R}^3} k^{\circ} a^+(k)\, a^-(k)\, dk, \qquad (k^{\circ} = (k^2 + m^2)^{1/2}).$

A priori (12) is quite doubtful from the mathematical viewpoint since it involves products of the distributions a^+ and a^-. However formulas (1) (2) - in which $\phi(x)$ is replaced by $(\delta(x - k)$ - give at once that

$$(13) \quad P_o \begin{pmatrix} K_o \\ K_1(x) \\ \cdot \\ \cdot \\ \cdot \\ K_n(x_1,...,x_n) \\ \cdot \\ \cdot \\ \cdot \end{pmatrix} = \begin{pmatrix} 0 \\ x^0 K_1(x) \\ \cdot \\ \cdot \\ \cdot \\ (x_1^0 + .. + x_n^0) \, K_n(x_1,...,x_n) \\ \cdot \\ \cdot \\ \cdot \end{pmatrix}$$

where $x_i^0 = (x_i^2 + m^2)^{\frac{1}{2}}$, and P_o is a well defined linear operator on the Fock space $\mathbb{F}$, with a dense

domain. We set, if $\theta \in \mathbb{R}$,

$$(14) \qquad\qquad\qquad U(\theta) = e^{i\theta P_o}$$

which is well defined as a unitary operator on $\mathbb{F}$ since one can prove P_o is self adfoint. One checks easily that

$$(15) \qquad\qquad\qquad A_o(x,t+\theta) = U(\theta) \, A_o(x,t) \, U(-\theta)$$

(same result with $\pi_o = \dfrac{\partial A_o}{\partial t}$ by differentiation in t).

Conclusion. If one drops some unessential points the theory of free fields makes sense within distribution theory. However this theory is in itself physically unuseful since if describes particles that do not interact. Its value comes from the fact that it is the preliminary stage of interacting field theory.

§7.3 THE INTERACTING FIELD EQUATION AND ITS SOLUTION.

The simplest model of a self interacting Boson field is ruled by the equation (initial value problem at time $t = \tau$)

$$(16) \quad \begin{cases} \left(-\dfrac{\partial^2}{\partial t^2} + \sum_{\mu=1}^{3} \dfrac{\partial^2}{\partial x_\mu^2} - m^2 \right) A(x,t) = g \, A^3(x,t) \\ A(x,\tau) = A_o(x,\tau) \\ (\dfrac{\partial A}{\partial t})(x, \tau) = (\dfrac{\partial A_o}{\partial t}) \, (x,\tau) \end{cases}$$

where A_0 is the free field operator (4); g is a real constant called the coupling constant ; if g = 0 one gets the free field.

All studies have shown that there is no reason to believe that the solution of (16) would have better mathematical properties than the free field (case g = 0 in (16)). Thus the term A^3 can be considered as irremediably meaningless within distribution theory. Now we are going to solve (16) by formal explicit calculations - that do not make sense within distribution theory.

Remark. (16) follows from the Lagrangian density

$$(17) \quad \mathcal{L}(x,t) = -\frac{1}{2}\left\{ \sum_{\mu=1}^{3} \left(\frac{\partial A}{\partial x_\mu}\right)^2 - \left(\frac{\partial A}{\partial t}\right)^2 + m^2 A^2 + \frac{g}{2} A^4 \right\} (x,t)$$

when one applies the principle of stationary action, see §7.2. (17) involves various undefined powers of the interacting field A.

For convenience we rewrite (16) in the form of a first order system in the t variable :

$$(16') \quad \begin{cases} \frac{\partial}{\partial t} A(x,t) = \pi(x,t) \\ \frac{\partial}{\partial t} \pi(x,t) = \sum_{\mu=1}^{3} \frac{\partial^2}{\partial x_\mu^2} A(x,t) - m^2 A(x,t) - g A(x,t)^3 \\ A(x,\tau) = A_0(x,\tau) \text{ and } \pi(x,\tau) = \pi_0(x,\tau) \end{cases} .$$

The first step in the resolution of (16) (16') is to replace it by an – a priori - more complicated system of equations ; but one will be able to solve explicitely this more complicated system. This method is called the "canonical Hamiltonian formalism". It consists in the following system of equations : (x,x' $\in \mathbb{R}^3, t \in \mathbb{R}$)

$$(18\ a) \begin{cases} [A(x,t)\ A(x',t')] = 0 = [\pi(x,t)\pi(x',t)] \\ [A(x,t),\ \pi(x',t)] = i\ \delta(x-x')\ Id \end{cases}$$

$$(18\ b) \begin{cases} \frac{\partial}{\partial t} A(x,t) = i \int_{\xi \in \mathbb{R}^3} [\mathcal{H}(\xi,t),\ A(x,t)]d\xi \\ \frac{\partial}{\partial t} \pi(x,t) = i \int_{\xi \in \mathbb{R}^3} [\mathcal{H}(\xi,t),\ \pi(x,t)]d\xi \end{cases}$$

where

(18 c) $\mathcal{H}(\xi,t) = \frac{1}{2}(\pi(\xi,t))^2 + \frac{1}{2}\sum_{\mu=1}^{3}(\frac{\partial}{\partial x_\mu})A(\xi,t)^2 + \frac{m^2}{2}(A(\xi,t))^2 + \frac{g}{4}(A(\xi,t))^4,$

with the initial condition

(18 d) $A(x,\tau) = A_0(x,\tau)$ and $\pi(x,\tau) = \pi_0(x,\tau).$

Proposition. Any solution (A,π) of (18 abcd) is necessarily also a solution of (16').

Of course this proposition has to be understood in a purely formal sense : it means that formal calculations (done by mimicking the classical formalism of integral and differential calculus) on (18 abcd) give (16'). These calculations run as follows :

From (18c) and (18a)

$[\mathcal{H}(\xi,t), A(x,t)] = \frac{1}{2}[(\pi(\xi,t))^2, A(x,t)] =$

$= \frac{1}{2}\{\pi(\xi,t)[\pi(\xi,t), A(x,t)] + [\pi(\xi,t), A(x,t)]\pi(\xi,t)\} =$

$= -i\,\delta(x-\xi)\,\pi(\xi,t)$

and

$[\mathcal{H}(\xi,t), \pi(x,t)] = \frac{1}{2}\left\{\left[\sum_{\mu=1}^{3}(\frac{\partial}{\partial x_\mu}A(\xi,t))^2, \pi(x,t)\right] + m^2[(A(\xi,t))^2, \pi(x,t)] + \frac{g}{2}[(A(\xi,t))^4, \pi(x,t)]\right.$

$= \sum_{\mu=1}^{3} i\frac{\partial}{\partial x_\mu}\,\delta(\xi-x).\frac{\partial}{\partial x_\mu}A(\xi,t) + i\,m^2\,\delta(\xi-x)\,A(\xi,t) + ig\,\delta(\xi-x)\,(A(\xi,t))^3.$

Therefore (18b) gives

$\frac{\partial}{\partial t}A(x,t) = \int_{\xi \in \mathbb{R}^3}\delta(x-\xi)\,\pi(\xi,t)\,d\xi = \pi(x,t)$

and

$$\frac{\partial}{\partial t}\,\pi\,(x,t) = -\left\{\sum_{\mu=1}^{3}\ \int \frac{\partial}{\partial x_\mu}\,\delta(\xi-x)\,\frac{\partial}{\partial x_\mu}\,A(\xi,t)\,d\xi\ +\right.$$

$$\left. +m^2 \int \delta(\xi-x)\,A(\xi,t)\,d\xi\ +\ g\ \int \delta(\xi-x)(A(\xi,t))^3\,d\xi\right\};$$

integration by parts in the first integral, and integrations of the δ functions give

$$\frac{\partial}{\partial t}\,\pi(x,t) = \sum_{\mu=1}^{3} \frac{\partial^2}{\partial x_\mu^{\,2}}\,A(x,t) - m^2\,A(x,t) - g\,(A(x,t))^3.\ \square$$

Now our task is to find an explicit solution to the Hamiltonian formalism (18abcd). This solution will be constructed from the free field operators $A_0(x,t)$ and $\pi_0(x,t)$. We set

$$(19)\quad \mathcal{H}_0(\xi,t) = \frac{1}{2}\,(\pi_0(\xi,t))^2 + \frac{1}{2}\sum_{\mu=1}^{3}\left(\frac{\partial}{\partial x_\mu}\,A_0(\xi,t)\right)^2 + \frac{m^2}{2}\,(A_0(\xi,t))^2 + \frac{g}{4}\,(A_0(\xi,t))^4$$

i. e. (18c) with A and π replaced by A_0 and π_0 respectively. Since we know that π_0 and A_0 are distributions whose powers are meaningless within distribution theory, so is (19). Then we set

$$(20)\quad H_0(t) = \int_{\xi \in \mathbb{R}^3} \mathcal{H}_0\,(\xi,t)\,d\xi.$$

Since $\mathcal{H}_0\,(\xi,t)$ itself is not defined mathematically it is impossible to discuss whether this integral makes sense. With 0 representing A or π, while 0_0 represents A_0 or π_0 respectively, we set

$$(21)\quad 0(x,t) = e^{i(t-\tau)\,H_0(\tau)}.\,0_0\,(x,\tau).e^{-i(t-\tau)H_0(\tau)}.$$

Let us check that $0(=A$ and $\pi)$ is solution of (18 abcd). (18a) follows at once from (9) by simplification of the exponentials. Putting $t = \tau$ in (21) gives (18 d). Putting (21) into (18c) gives (simplification of exponentials and (19))

$$\mathcal{H}\,(\xi,t) = e^{i(t-\tau)H_0(\tau)}\,\mathcal{H}_0(\xi,\tau)\,e^{-i(t-\tau)H_0(\tau)}.$$

Integration in ξ and (20) give

$$\int \mathcal{H}(\xi,t)\,d\xi\ = e^{i(t-\tau)H_0(\tau)}\,H_0(\tau)\,e^{-i(t-\tau)H_0(\tau)}.$$

Simplification of the exponentials (usually the exponential of an operator commutes with this operator) gives

(22) $\int \mathcal{H}(\xi,t)\, d\xi = H_0(\tau).$

Now let us differentiate (21) in time :

$$\frac{d}{dt}\, 0(x,t) = i\, H_0(\tau)\, 0(x,t) - i\, 0(x,t)\, H_0(\tau)$$

$$= i\, [H_0(\tau),\, 0(x,t)]$$

$$= i \int [\mathcal{H}(\xi,t),\, 0(x,t)]\, d\xi \ \text{from} \ (22).$$

Conclusion : (21) is an explicit solution of the interacting field equation (16) : we have calculated the interacting field operators ! But this has been done at the price of a great deal of calculations that are mathematically meaningless (as far as one knows, in particular within distribution theory).

Problem. Try to solve scalar equations (i. e. whose unknowns are scalar valued) by explicit similar calculations. The scalar case being much clearer (than the present case in which the unknowns are operators on an infinite dimensional Hilbert space) similar calculations could perhaps be reproduced rigorously in $\mathcal{G}$, and shed some light on the calculations in this section, and also on the way to solve PDEs in $\mathcal{G}$. See explicit calculations in Colombeau-Oberguggenberger [2], Oberguggenberger [3, 4, 10].

§7. 4 THE SCATTERING OPERATOR AND THE RESULTS OF THE THEORY.

The predictive results of the theory are obtained from the "scattering operator" defined as follows : with $U(\theta)$ defined in (14) we set

(23) $S_\tau(t) = U(t-\tau)e^{-i(t-\tau)H_0(\tau)};$

from (21) and (15,14)

(24) $0(x,t) = (S_\tau(t))^{-1}\, 0_0(x,t)\, S_\tau(t).$

Setting $t = +\infty$ and $\tau = -\infty$ (i. e. some kind of limit that we are unable to study since we work on objects that are not well defined mathematically) (24) becomes

(25) $0(x, +\infty) = S^{-1}\, 0_0(x,+\infty)\, S$

if

(26) $S = S_{-\infty}(+\infty).$

S is considered as a unitary operator on $\mathbb{F}$ since the operators $S_\tau(t)$ are considered as such ; it is called the scattering operator and it depends on the coupling constant g (S = Id if g = 0) ; so it will be written S(g) in the sequel. Imagine that particles collide : we can imagine that before the collision(roughly speaking in the remote past i. e. at $t = -\infty$) they were free and governed by the free field equation ; we can also imagine that in the future of the collision the particles created in the collision are also free : they are governed by $0(x,+\infty)$ since collision has taken place.

Assume that an observable is a finite product $\prod\limits_{i=1}^{q} 0_i(x,+\infty)$ where $0_i = 0$ or partial derivatives of 0. Let K_1, K_2 be two elements of the Fock space. Then the expectation value

$$< K_1, \left(\prod_{i=1}^{q} 0_i\right) K>_{\mathbb{F}}$$

is a complex number whose absolute value can be interpreted as physically meaningful. Let $0_{i,0}$ be the corresponding objects obtained with 0_0 in place of 0. From (25)

$$< K_1, \left(\prod_{i=1}^{q} 0_i\right) K_2 >_{\mathbb{F}} = < K_1, \left(S^{-1}\prod_{i=1}^{q} 0_{i,0}\ S\right) K_2 >_{\mathbb{F}}$$

$$= < S\,K_1, \left(\prod_{i=1}^{q} 0_{i,0}\right) S\,K_2 >_{\mathbb{F}}$$

since S is assumed to be a unitary operator. Since the 0_i's and $0_{i,0}$'s there are considered at the time t $= +\infty$ then SK_1 and SK_2 are the states of a free field at t $= +\infty$ giving the same expectation values as the states K_1 and K_2 that have interacted. The numerical results are transition probabilities from a given state at $-\infty$ (i. e. before interaction) to another given state at $+\infty$ (i.e. after interaction). From the above explanations they are obtained via the scattering operator, more precisely from scalar products $<K_1, SK_2 >_{\mathbb{F}}$.

Now we are interested in an explicit calculation of $S_\tau(t)$ and S. From (23,14)

$$\frac{d}{dt} S_\tau(t) = iP_0\, e^{i(t-\tau)P_0} e^{-i(t-\tau)H_0(\tau)} -i\, e^{i(t-\tau)P_0}\, H_0(\tau) e^{-i(t-\tau)H_0(\tau)}$$

i. e

$$i\frac{d}{dt} S_\tau(t) = -P_0 U(t-\tau) e^{-i(t-\tau)H_0(\tau)} + U(t-\tau)\, H_0(\tau) e^{-i(t-\tau)H_0(\tau)}$$

$$= \{-P_0 + U(t-\tau)\, H_0(\tau)\, U\,(t-\tau)^{-1}\}\, U(t-\tau) e^{-i(t-\tau)H_0(\tau)}.$$

Using (23) for the last two factors and (15,19,20) for the second term in the first factor one obtains

$$(27) \quad i\frac{d}{dt} S_\tau(t) = (-P_0 + H_0(t))\, S_\tau(t).$$

From (20) (19) (4) (1)(2) one can compute explicitely $H_0(t)$.

This explicit calculation is purely formal - since it involves meaningless products of distributions - and (although cumbersome : it takes some two pages) it can be done easily by the reader. It is done in detail in Kastler [1] and Colombeau [2] p 22 - 24. One finds

$$(28) \quad H_0(t) = P_0 + \frac{1}{2}(2\pi)^{-3} \int_{k \in \mathbb{R}^3} k^\circ dk + \frac{g}{4} \int_{\xi \in \mathbb{R}^3} (A_0(\xi,t))^4 \, d\xi.$$

Looking more carefully at this expression we notice at once that the integral $\int_{k \in \mathbb{R}^3} k^\circ dk$ is divergent

(recall $k^\circ = (k^2 + m^2)^{1/2}$) ! This is not so amazing if one has in mind that most calculations leading to (28) are mathematically forbidden. Physicists suppress purely and simply this undesirable infinite quantity and so they state (28) as

$$(28') \quad H_0(t) = P_0 + \frac{g}{4} \int_{\xi \in \mathbb{R}^3} (A_0(\xi,t))^4 \, d\xi.$$

Therefore (27) becomes

$$(29) \quad \frac{d}{dt} S_\tau(t) = -i \frac{g}{4} \int_{\xi \in \mathbb{R}^3} (A_0(\xi,t))^4 \, d\xi \cdot S_\tau(t).$$

Since furthermore it is obvious that $S_\tau(\tau) = \text{Id}$, then $S_\tau(t)$ is governed by an "ordinary" linear differential equation (not so much ordinary since A_0^4 does not make sense). Developing formally the solution in powers of g one obtains at once

$$(30) \quad S_\tau(t) = \text{Id} + \sum_{n=1}^{+\infty} S_\tau^{(n)}(t)$$

$$(31) \quad S_\tau^{(n)}(t) = (-i)^n \left(\frac{g}{4}\right)^n \int_{\substack{\tau \leq \lambda_1 \leq t \\ \tau \leq \lambda_2 \leq \lambda_1 \\ \cdot \\ \cdot \\ \cdot \\ \tau \leq \lambda_n \leq \lambda_{n-1}}} \int_{\substack{\xi_i \in \mathbb{R}^3 \\ 1 \leq i \leq n}} (A_0(\xi_1,\lambda_1))^4 ... (A_0(\xi_n,\lambda_n))^4 \, d\xi_1...d\xi_n \, d\lambda_1...d\lambda_n.$$

For $\tau = -\infty$ and $t = +\infty$ one obtains

$$(32) \quad S = \text{Id} + \sum_{n=1}^{+\infty} S_n$$

(33) $\quad S_n=(-i)^n\left(\frac{g}{4}\right)^n \underset{\substack{x_i \in \mathbb{R}^3 \\ t_i \in \mathbb{R} \\ 1\leq i\leq n}}{\int} Y(t_1-t_2)...Y(t_{n-1}-t_n)\,(A_0(x_1,t_1))^4...(A_0(x_n,t_n))^4\,dx_1...dx_n\,dt_1...dt_n.$

where Y is the Heaviside function. In Quantum electrodynamics the coupling constant (playing the role of g) is "small" $\left(\frac{1}{137}\right)$ and so one has the idea to approximate S by the sum of a few first terms in the power series development (32), i. e. to compute for instance $< K_1, (\mathrm{Id} + S_1 + S_2)\,K_2 >$; this gives results in good agreement with those from experiments, see Kastler [1] for some classical situations in Quantum Electrodynamics. But some terms in these calculations appear in the form of divergent integrals (of a nature different from the one of the infinite quantity met in (28)) ; for instance, after some calculations (see Bogoliubov - Shirkov [1]) one finds integrals of the kind

(34) $\qquad \underset{\bar{q}\in \mathbb{R}^4}{\int} \dfrac{1}{(\bar{q}^2 + m^2 - i\eta)\{(\bar{p}-\bar{q})^2 + m^2 - i\eta\}}\,d\bar{q}$

where $\bar{q} = (q_1, q_2, q_3, q') \in \mathbb{R}^4$, $\bar{p} = (p_1, p_2, p_3, p') \in \mathbb{R}^4$, $\bar{q}^2 = q_1^2 + q_2^2 + q_2^3 - q'^2$, $\eta > 0$. The problem is that such integrals are not convergent (in spherical coordinates (34) is of the kind

$\int_\alpha^{+\infty} \dfrac{\rho^3 d\rho}{\rho^4}$ i. e. $\int_\alpha^{+\infty} \dfrac{d\rho}{\rho}$ at $+\infty$, $\alpha > 0$ unimportant)

Twenty years after the inception of the theory a procedure for extracting finite results from these divergent integrals has been discovered (1947). For (34) it consists in substracting the "infinite quantity" $\underset{\bar{q}\in \mathbb{R}^4}{\int} \dfrac{1}{(\bar{q}^2 + m^2 - i\eta)}\,d\bar{q}$

(value of the one in (33) at $\bar{p} = 0$) : this substraction gives a convergent integral. Besides a considerable amount of work this process, called Renormalization Theory, has never been completely clarified and looks as an ambiguous set of unjustified recipes.

Research Problem. General calculations looking like those of the interacting field theory are in principle meaningful in our setting of generalized functions. But the fact that the values of the basic objects (the free fields) are unbounded operators makes the analysis of these calculations very delicate. A very rough cleaning has been attempted in Colombeau [1] Part 3. The awkward presentation there (modification of the sets A_q to explain the removal of divergences) can be corrected (see the appendix of Colombeau [13] : instead of modifying the sets A_q it appears that the correct process is as follows : only the macroscopic aspect of the free field operators is known ; renormalization would consist in choosing them among the generalized functions that have this macroscopic aspect.). It would be very interesting to transform this rough attempt into a decent cleaning, that might perhaps be predictive, as this can be done for less technical theories such as those presented in chapters 3 to 6 of this book.

Chapter 8. The abstract theory of generalized functions

The aim of this chapter is to introduce the definition of $\mathcal{G}(\Omega)$ in a natural way . Another short introduction may be found in the survey Colombeau [14].

§8.1. THE IDEA TO MULTIPLY LINEAR FORMS AND TO USE A QUOTIENT.

Distributions are linear maps from the space $\mathcal{D}(\Omega)$ into $\mathbb{C}$. One might think of their products as nonlinear maps from the space $\mathcal{D}(\Omega)$ into $\mathbb{C}$: then if T_1, T_2 are distributions the product $T_1 T_2$ would be the map

$$\varphi \to T_1(\varphi) . T_2(\varphi)$$

which is a monomial of degree 2 on $\mathcal{D}(\Omega)$. If δ is the Dirac delta function then its square δ^2 would then be defined as the map $\varphi \to (\varphi(0))^2$. However let f_1, f_2 be two $\mathcal{C}^\infty$ functions on Ω. When considered as distributions they are the linear forms

$$\varphi \to \int_\Omega f_i(x) \, \varphi(x) \, dx \qquad i = 1,2.$$

Their product as linear forms is thus the map

(1)
$$\varphi \to \int_\Omega f_1(x) \, \varphi(x) \, dx . \int_\Omega f_2(x) \, \varphi(x) \, dx$$

while their classical product $f_1 \, f_2 \in \mathcal{C}^\infty(\Omega)$, when considered as a distribution, is the map

(2)
$$\varphi \to \int f_1(x) \, f_2(x) \, \varphi(x) \, dx.$$

The right hand side members of (1) and (2) are different in general, so the product of linear forms appears to be incoherent even with the product of $\mathcal{C}^\infty$ functions. A solution to this incoherence arose from more abstract considerations (Colombeau [2, 5, 6]) and § 8.2 but the resulting idea is quite simple : <u>identify the two right hand sides of (1) and (2) by means of a quotient.</u> So let us consider a large enough space in which one can define all product maps $\varphi \to \prod_{i=1}^{m} T_i(\varphi)$, $T_i \in \mathcal{D}'(\Omega)$, as well as many nonlinear operations $\varphi \to f(T_1(\varphi),...,T_m(\varphi))$, $f \in \mathcal{C}^\infty(\mathbb{R}^{2m})$ [note that for the applications it follows clearly from chap. 2 and 7 that we need much more than products of distributions]. This large enough space might be the space $\mathcal{C}^\infty(\mathcal{D}(\Omega))$ of all complex valued $\mathcal{C}^\infty$ functions from $\mathcal{D}(\Omega)$ into $\mathbb{C}$.

Since this chapter is aimed at the mathematician reader let us say a few words on that : the space $\mathcal{D}(\Omega)$ is equipped with a natural structure of locally convex topological vector space (a structure generalizing the one of normed spaces) and the distributions are the linear continuous forms on $\mathcal{D}(\Omega)$, i.e. $\mathcal{D}'(\Omega)$ is the topological dual of $\mathcal{D}(\Omega)$. The mathematician reader knows the concept of

$\mathscr{C}^\infty$ functions defined on a normed space. The case of $\mathscr{C}^\infty$ functions defined on a locally convex space is a straightforward extension (provided one adopts the proper definition, see Colombeau [1]). The reader may consider, without any trouble, that one manipulates them like in the normed space case. If $f \in \mathscr{C}^\infty(\mathbb{R}^{2m})$, if $T_1, ..., T_m \in \mathscr{D}'(\Omega)$ then the map

$$\varphi \to f(T_1(\varphi),...,T_m(\varphi))$$

is in $\mathscr{C}^\infty(\mathscr{D}(\Omega))$. Thus $\mathscr{C}^\infty(\mathscr{D}(\Omega))$ appears as large enough to contain all "useful" nonlinear functions of distributions.

In order to identify (1) and (2) which are different elements in $\mathscr{C}^\infty(\mathscr{D}(\Omega))$ one should consider a quotient of $\mathscr{C}^\infty(\mathscr{D}(\Omega))$. This quotient was discovered in the following way.

§ 8.2. DETERMINATION OF THE QUOTIENT.

We denote by $\mathscr{E}(\Omega)$ the vector space of all complex valued C^∞ functions on Ω, equipped with its usual topology of uniform convergence on the compact subsets of Ω, (for functions and all partial derivatives) and by $\mathscr{E}'(\Omega)$ its topological dual. The reader aware of distribution theory knows that $\mathscr{E}'(\Omega)$ can be identified with the space of distributions with compact support in Ω. $\mathscr{E}'(\Omega)$ is naturally a locally convex space and so one can define the complex valued $\mathscr{C}^\infty$ functions on $\mathscr{E}'(\Omega)$; we denote their vector space by $\mathscr{C}^\infty(\mathscr{E}'(\Omega))$. From distribution theory one knows that $\mathscr{D}(\Omega)$ is everywhere dense in $\mathscr{E}'(\Omega)$, and so the restriction map

$$\mathscr{C}^\infty(\mathscr{E}'(\Omega)) \qquad \mathscr{C}^\infty(\mathscr{D}(\Omega))$$
$$\Phi \to \Phi \mid_{\mathscr{D}(\Omega)}$$

is injective. One notes $\mathscr{C}^\infty(\mathscr{E}'(\Omega)) \subset \mathscr{C}^\infty(\mathscr{D}(\Omega))$. Let δ_x denote the Dirac function centered at $x \in \Omega$ i.e. $\delta_x(\varphi) = \varphi(x)$ if $\varphi \in \mathscr{D}(\Omega)$. Of course $\delta_x \in \mathscr{E}'(\Omega)$. Let $\mathscr{M}$ be the set of all $\phi \in \mathscr{C}^\infty(\mathscr{E}'(\Omega))$ such that $\phi(\delta_x) = 0$ for all $x \in \Omega$. One proves easily that the quotient space $\dfrac{\mathscr{C}^\infty(\mathscr{E}'(\Omega))}{\mathscr{M}}$ is isomorphic to the space $\mathscr{E}(\Omega)$ of all $\mathscr{C}^\infty$ functions on Ω. This leads us to the idea : extend $\mathscr{M}$ to a certain subset $\mathscr{N}$ of $\mathscr{C}^\infty(\mathscr{D}(\Omega))$ and take the quotient $\dfrac{\mathscr{C}^\infty(\mathscr{D}(\Omega))}{\mathscr{N}}$ as our space of generalized functions : in this way, if $\mathscr{N} \cap \mathscr{C}^\infty(\mathscr{E}'(\Omega)) = \mathscr{M}$ then $\mathscr{E}(\Omega) = \dfrac{\mathscr{C}^\infty(\mathscr{E}'(\Omega))}{\mathscr{M}}$ would be exactly a subset of $\dfrac{\mathscr{C}^\infty(\mathscr{D}(\Omega))}{\mathscr{N}}$ and so, in this quotient, (1) and (2) would be identified. By definition $\mathscr{M} = \{\phi \in \mathscr{C}^\infty(\mathscr{E}'(\Omega))$ such that $\phi(\delta_x) = 0 \ \forall x \in \Omega\}$. But $\phi(\delta_x)$ does not make sense in the more general case $\phi \in \mathscr{C}^\infty(\mathscr{D}(\Omega))$. One should then approximate the Dirac function δ by functions δ^ε, $\delta^\varepsilon \in \mathscr{D}(\mathbb{R}^n)$, since, if $\delta_x^\varepsilon(\lambda) = \delta^\varepsilon(\lambda - x)$, then

$\delta_x^\varepsilon \to \delta_x$ when $\varepsilon \to 0$, and $\phi(\delta_x^\varepsilon)$ makes sense. This can be done as follows :

<u>Definition.</u> If $q = 0,1,2,...$ we set $\mathcal{A}_q = \{\varphi \in \mathcal{D}(\mathbb{R}^n)$ such that $\int \varphi(x)\,dx = 1$ and $\int x^i \varphi(x)\,dx = 0$ if $1 \le |i| \le q\}$ (here $i = (i_1,...,i_n) \in \mathbb{N}^n$ and $|i| = i_1 + ... + i_n$: usual multiindex notation).

<u>Proposition.</u> For any q the set $\mathcal{A}_q$ is non void. The sets $\mathcal{A}_q$ are decreasing and $\underset{q \in \mathbb{N}}{\cap}\, \mathcal{A}_q = \varnothing$.

The proof is left as an exercise to the mathematician reader (see Colombeau [2, 3] for instance). We set

$$(3) \qquad \varphi_{\varepsilon,x}(\lambda) = \frac{1}{\varepsilon^n}\,\varphi\left(\frac{\lambda - x}{\varepsilon}\right)$$

and $\varphi_\varepsilon = \varphi_{\varepsilon,0}$. Then $\varphi \in \mathcal{A}_q \Leftrightarrow \varphi_\varepsilon \in \mathcal{A}_q$ and $\varphi_\varepsilon \to \delta$ as $\varepsilon \to 0$, as soon as $\varphi \in \mathcal{A}_0$.

The following is a characterization of $\mathcal{M}$ in terms of the functions $\varphi_{\varepsilon,x}$ and so it is extendable to $\mathcal{C}^\infty(\mathcal{D}(\Omega))$:

<u>Proposition.</u> Let $\phi \in \mathcal{C}^\infty(\mathcal{E}'(\Omega))$. Then $\phi \in \mathcal{M}$ (i.e. $\phi(\delta_x) = 0\ \forall x \in \Omega$) if and only if $\forall q \in \mathbb{N},\ \forall\, \varphi \in \mathcal{A}_q,\ \forall K$ compact subset of $\Omega\ \exists\, c > 0$ and $\eta > 0$ such that

$$|\phi(\varphi_{\varepsilon,x})| \le c\varepsilon^{q+1} \text{ if } 0 < \varepsilon < \eta \text{ and } x \in K.$$

The proof is in Colombeau [2] prop. 3.3.3.

Let $\overline{\mathcal{M}} = \{\phi \in \mathcal{C}^\infty(\mathcal{D}(\Omega))$ such that $\forall q \in \mathbb{N},\ \forall \varphi \in \mathcal{A}_q,\ \forall K$ compact in $\Omega\ \exists\, c > 0$ and $\eta > 0$ such that $|\,\phi(\varphi_{\varepsilon,x})| \le c\varepsilon^{q+1}$ if $0 < \varepsilon < \eta$ and $x \in K\}$. Is $\overline{\mathcal{M}}$ an ideal of $\mathcal{C}^\infty(\mathcal{D}(\Omega))$? (the word "ideal" means essentially that $\forall\, \phi \in \mathcal{C}^\infty(\mathcal{D}(\Omega)),\ \forall\, \psi \in \overline{\mathcal{M}}$ we have $\phi\,\psi \in \overline{\mathcal{M}}$; this property would be essential in order to define the product in the quotient space (modulo $\overline{\mathcal{M}}$)). One can readily check that the answer is no : the function $\phi : \varphi \to \exp(\varphi(0))$ is such that $\phi(\varphi_\varepsilon) = \exp\left(\frac{1}{\varepsilon^n}\,\varphi(0)\right)$ grows too fast when $\varepsilon \to 0$ (if $\varphi(0) \ne 0$) : $\mathcal{C}^\infty(\mathcal{D}(\Omega))$ is too large ! But it suffices of minor modifications in definitions to repair that.

<u>Exercise.</u> Prove that the difference of the two right hand side members of (1) and (2) is in $\overline{\mathcal{M}}$ (state this difference as

$$\int f_1(x+\varepsilon\mu)\,\varphi(\mu)\,d\mu.\ \int f_2(x+\varepsilon\mu)\,\varphi(\mu)\,d\mu - \int f_1(x+\varepsilon\mu)\ f_2(x+\varepsilon\mu)\,\varphi(\mu)\,d\mu$$

and use Taylor's formula up to order q if $\varphi \in \mathcal{A}_q$)

§8.3 CONSTRUCTION OF AN ALGEBRA g(Ω) We consider the set S of all $\mathcal{C}^\infty$ functions ϕ on $\mathcal{D}(\Omega)$ such that $|\phi(\phi_{\varepsilon,x})|$ is bounded above by a power of $\frac{1}{\varepsilon}$ when $\varepsilon \to 0$, independently of $\phi \in \mathcal{A}_q$, for large enough q, and independently of x when x ranges in any compact subset of Ω. This will permit, by defining a quotient of the type $\dfrac{S}{\overline{\mathcal{M}}}$, an automatic definition of the product of two

elements in this quotient (since $\overline{\mathcal{M}}$ would be an ideal in S). But partial x - derivatives of two elements of $\dfrac{S}{\overline{\mathcal{M}}}$ would not be definable in general. So we modify the definition of S and $\overline{\mathcal{M}}$ so as

to have the requested properties for any partial derivative. This leads at once to the following definitions.

Definition 1. If $\phi \in \mathcal{C}^\infty (\mathcal{D}(\Omega))$ we say that ϕ is moderate if for every compact subset K of Ω and every partial derivation

$$D = \frac{\partial^{|k|}}{\partial x_1^{k_1}...\partial x_n^{k_n}} \qquad (0 \le |k| < +\infty)$$

there is an $N \in \mathbf{N}$ such that
$$\forall \phi \in \mathcal{A}_q, \text{ with q large enough, } \exists \, c > 0 \text{ and } \eta > 0 \text{ such that}$$

$$|D\phi\,(\phi_{\varepsilon,x})| \le c(\tfrac{1}{\varepsilon})^N$$

if $x \in K$ and $0 < \varepsilon < \eta$.

We denote by $\mathcal{C}_M^\infty(\mathcal{D}(\Omega))$ or $\mathcal{E}_M(\mathcal{D}(\Omega))$ the set of all moderate elements of $\mathcal{C}^\infty (\mathcal{D}(\Omega))$. One has the

three properties

      a) $\mathcal{E}_M(\mathcal{D}(\Omega))$ is an algebra (namely $\phi_1. \phi_2 \in \mathcal{E}_M(\mathcal{D}(\Omega))$ if $\phi_1, \phi_2 \in \mathcal{E}_M(\mathcal{D}(\Omega))$: use Leibniz's formula)

      b) $D\,\phi \in \mathcal{E}_M(\mathcal{D}(\Omega))$ if $\phi \in \mathcal{E}_M(\mathcal{D}(\Omega))$ (obvious)

      c) $\mathcal{D}'(\Omega) \subset \mathcal{E}_M(\mathcal{D}(\Omega))$ (easy consequence of the structure theorem recalled in § 1.3).

Definition 2. We set $\mathcal{N}(\Omega) = \{ \phi \in \mathcal{E}_M(\mathcal{D}(\Omega))$ such that for every K and D (as in def.1) there is an N $\in \mathbf{N}$ such that

$\forall \varphi \in \mathfrak{K}_q$, with q large enough, $\exists\, c > 0$ and $\eta > 0$ such that

$$|D\phi\,(\varphi_{\varepsilon,x})| \le c\,(\varepsilon)^{q-N}$$

if $x \in K$ and $0 < \varepsilon < \eta\}$.

One has the three properties
  a) $\mathcal{N}(\Omega)$ is an ideal of the algebra $\mathcal{E}_M(\mathcal{D}(\Omega))$ (obvious)
  b) $D\phi \in \mathcal{N}(\Omega)$ if $\phi \in \mathcal{N}(\Omega)$ (obvious)
  c) $\mathcal{D}'(\Omega) \cap \mathcal{N}(\Omega) = \{0\}$ (see Colombeau [2] prop.3.4.7)
We are now ready to define our quotient :

<u>Definition 3</u> .We set

$$g(\Omega) = \frac{\mathcal{E}_M(\mathcal{D}(\Omega))}{\mathcal{N}(\Omega)}.$$

The elements of $g(\Omega)$ have most of the properties listed in chap.3.

One understands how our construction is natural, since it follows rather simple and even standard mathematical reasoning. However, these objects have the major defect that their construction is not elementary, and so not accessible to the majority of applied mathematicians, physicists and engineers. Fortunately it can be made elementary by some straightforward simplifications : the concept of differentiability over the locally convex space $\mathcal{D}(\Omega)$ is unessential. It emerged from the concept of distribution as a linear continuous map (and so differentiable) on $\mathcal{D}(\Omega)$. But since only differentiation in the x variable is explicitely used in the definitions, one can drop the differentiability in the variable $\varphi \in \mathcal{D}(\Omega)$. This leads to an elementary definition that we expose in the next section.

<u>Remarks :</u> The concept of association can be easily defined in $g(\Omega)$, see Colombeau [1] §3.5 ; we shall define it in the next chapter. Through the association the multiplication in $g(\Omega)$ realizes a synthesis of most existing particular products of distributions, see Colombeau [2] §3.5, Rosinger [1], Oberguggenberger [1], Colombeau - Oberguggenberger [1], Biagioni [1].

§8.4 <u>AN ELEMENTARY THEORY OF GENERALIZED FUNCTIONS.</u>
  The presentation given here is the one in Colombeau [3] Aragona-Colombeau [1], exposed also in part II of Rosinger [1]. An improvement (but at the price of being slightly more technical) is given in Aragona-Biagioni [1], Biagioni [1]. Several slight modifications can also be proposed (Biagioni [1] § 1.10 for instance). Let Ω be an open set in $\mathbb{R}^n$; we define the space $\mathcal{G}(\Omega)$ (slightly different from the space $g(\Omega)$ of §8.3) as follows.

<u>Notation</u> We denote by $\mathscr{E}[\Omega]$ the set of all functions R : $\mathscr{A}_0 \times \Omega \to \mathbb{C}$

$$\varphi, x \qquad\qquad R(\varphi, x)$$

which are $\mathscr{C}^\infty$ in the variable $x \in \Omega$, for any given φ.

We denote by $\mathscr{E}_M[\Omega]$ the set of all $R \in \mathscr{E}[\Omega]$ such that $\forall K$ compact in Ω $\forall D$ partial derivative $\exists N \in \mathbb{N}$ such that

$\forall \varphi \in \mathscr{A}_q($ with q large enough$)$ $\exists c > 0, \eta > 0$ such that

$$|DR(\varphi_\varepsilon, x)| \le c(\frac{1}{\varepsilon})^N$$

if $x \in K$ and $0 < \varepsilon < \eta$.

We denote by Γ the set of all increasing functions $\gamma : \mathbb{N} \to \mathbb{R}^+$ such that $\gamma(q) \to +\infty$ when $q \to +\infty$.

We denote by $\mathscr{N}[\Omega]$ the set of all $R \in \mathscr{E}_M[\Omega]$ such that $\forall K, \forall D$ $\exists \gamma \in \Gamma$ such that if $\varphi \in \mathscr{A}_q($with q large enough$)$ $\exists c > 0, \eta > 0$ such that

$$|DR(\varphi_\varepsilon, x)| \le c\, e^{\gamma(q)}$$

if $x \in K$ and $0 < \varepsilon < \eta$.

Finally we set

$$\mathscr{G}(\Omega) = \frac{\mathscr{E}_M[\Omega]}{\mathscr{N}[\Omega]},$$

that is, an element $G \in \mathscr{G}(\Omega)$ is an equivalence class $R + \mathscr{N}[\Omega]$ of an element $R \in \mathscr{E}_M[\Omega]$. Since $\mathscr{N}[\Omega]$ is an ideal of the algebra $\mathscr{E}_M[\Omega]$, the product $G_1 G_2$ is defined as the class of $R_1 R_2$ if R_i, $i = 1,2$, is a (arbitrary) representative of G_i ; similarly DG is the class of DR if D is any partial derivative operator. In the classical construction of $\mathbb{R}$ from $\mathbb{Q}$ by the method of Cauchy sequences the sets $\mathscr{E}[\Omega]$, $\mathscr{E}_M[\Omega]$ and $\mathscr{N}[\Omega]$ are respectively replaced by the sets of all sequences, all Cauchy sequences and all sequences convergent to 0 of rational numbers.

<u>The inclusion $\mathscr{C}^\infty(\Omega) \subset \mathscr{G}(\Omega)$.</u> To $f \in \mathscr{C}^\infty(\Omega)$ we associate the class of $R(\varphi, x) = f(x)$.

<u>The inclusion $\mathscr{E}(\Omega) \subset \mathscr{G}(\Omega)$.</u> To $f \in \mathscr{C}(\Omega)$ we associate the class of $R(\varphi, x) = \int f(\lambda)\, \varphi(\lambda - x)\, d\lambda$

when the function $\lambda \to \varphi(\lambda - x)$is in $\mathscr{D}(\Omega)$ (slight complication if not, resolved in a standard way). The same formula gives an inclusion of $L^p(\Omega)$, $1 \le p \le +\infty$, into $\mathscr{G}(\Omega)$.

Exercise. If f $\in$ $\mathscr{C}^\infty(\Omega)$ prove that the map

$$(\varphi,x) \to \int f(\lambda) \; \varphi(\lambda-x) \; d\lambda - \varphi(x)$$

is in $\mathscr{N}[\Omega]$ (thus the above inclusions are coherent with the classical inclusion $\mathscr{C}^\infty(\Omega) \subset \mathscr{C}(\Omega)$).

The inclusion $\mathscr{D}'(\Omega) \subset \mathscr{G}(\Omega)$. To $T \in \mathscr{D}'(\Omega)$ we associate the class of $R(\varphi,x) = (T * \overset{\vee}{\varphi})(x)$ if $\overset{\vee}{\varphi}(\lambda) = \varphi(-\lambda)$. One can show easily that $R \in \mathscr{E}_M[\Omega]$ (use for instance the local structure theorem of distributions, see §1.3).

The association is defined as follows : $G \in \mathscr{G}(\Omega)$ is said to be associated to 0 iff :
$$\forall \, \psi \in \mathscr{D}(\Omega) \; \exists \, N \in \mathbb{N} \text{ such that } \varphi \in \mathscr{R}_N \Rightarrow$$

$$\int_\Omega R(\varphi_\varepsilon, x) \; \psi \; (x) \; dx \to 0 \text{ when } \varepsilon \to 0.$$

(if R is an arbitrary representative of G : if the above holds for some representative then it holds for any representative of G). $G_1, G_2 \in \mathscr{G}(\Omega)$ are said to be associated iff $G_1 - G_2$ is associated to 0.

Example. If H is a Heaviside generalized function defined as in 3.3.1 (note : the definition of the restriction of $G \in \mathscr{G}(\Omega)$ to $\omega \subset \Omega$, ω open, is obvious ; c) in 3.3.1 can be stated as : $\forall \, \varphi \in \mathscr{R}_q$ (q large enough) $\exists \, \eta > 0$ such that $\underset{\substack{0<\varepsilon<\eta \\ -1\le x\le 1}}{\sup} \; |R(\varphi_\varepsilon,x)| < +\infty$) then

$$H^n \approx H \qquad \forall n = 1,2,3,...$$

and $H^n \ne H$ if $n \ne 1$.

The definition of $\mathscr{G}(\Omega)$ is fully used for some applications that have a theoretical character such as a synthesis of most existing multiplications of distributions or general existence-uniqueness results for PDEs. But for the applications considered in this book a simplified presentation is sufficient. It gives a simplified algebra $\mathscr{G}_s(\Omega)$ - the subscript s stands for "simplified" - which is a subalgebra of $\mathscr{G}(\Omega)$ but which does not contain canonically $\mathscr{D}'(\Omega)$. Thus the respective situation of these spaces is

$$\mathscr{C}^\infty(\Omega) \begin{array}{c} \subset \mathscr{D}'(\Omega) \subset \\[4pt] \\[4pt] \subset \mathscr{G}_s(\Omega) \subset \end{array} \mathscr{G}(\Omega)$$

$\mathcal{G}_s(\Omega)$ and $\mathcal{G}(\Omega)$ are differential algebras of generalized functions which can be equally well described by §3.1 and §3.2 . But there is no canonical inclusion of the set $\mathcal{D}'(\Omega)$ of all distributions on Ω into $\mathcal{G}_s(\Omega)$ so that the multiplication in $\mathcal{G}_s(\Omega)$ does not provide a canonical multiplication of distributions.

Definition of the "simplified" differential algebra $\mathcal{G}_s(\Omega)$

We denote by $\mathcal{E}_s[\Omega]$ the set of the maps

$R :]0,1] \times \Omega \to \mathbb{C}$ which are C^∞ in the variable x. Obviously, if

$$D = \frac{\partial^{k_1 + \ldots + k_n}}{\partial x_1^{k_1} \ldots \partial x_n^{k_n}} \ , \ k_i \in \mathbb{N},$$ is any partial derivative operator, then DR is in $\mathcal{E}_s[\Omega]$ if R is in $\mathcal{E}_s[\Omega]$.

If K is a compact subset of Ω (i.e. K is bounded, closed, and lies at a strictly positive distance from the boundary of Ω) then we set

$$f_{K,D}(R,\varepsilon) = \sup_{x \in K} |DR(\varepsilon,x)|.$$

We denote by $\mathcal{E}_{M,s}[\Omega]$ the subset of $\mathcal{E}_s[\Omega]$ of the maps R for which, for every K and D, $f_{K,D}(R,\varepsilon)$ is bounded above by a power of $\frac{1}{\varepsilon}$ when $\varepsilon \to 0$. Example : n = 1, $R(\varepsilon,x) = \frac{1}{\varepsilon} \rho\left(\frac{x}{\varepsilon}\right)$ for fixed $\rho \in \mathcal{D}(\mathbb{R})$.

We denote by $\mathcal{N}_s[\Omega]$ the subset of $\mathcal{E}_{M,s}[\Omega]$ of the maps R for which, for every K and D, $f_{K,D}(R,\varepsilon)$ tends to 0 (when $\varepsilon \to 0$) faster than any power of ε (one says that $f_{K,D}(R,\varepsilon)$ decreases rapidly to 0 when $\varepsilon \to 0$). Example : n = 1, $\rho \in \mathcal{D}(\mathbb{R})$, $R(\varepsilon,x) = e^{-1/\varepsilon}\rho\left(\frac{x}{\varepsilon}\right)$.

If $R \in \mathcal{E}_{M,s}[\Omega]$ then obviously $DR \in \mathcal{E}_{M,s}[\Omega]$. $\mathcal{E}_{M,s}[\Omega]$ is a vector space and the pointwise product $R_1 R_2$ is an element of $\mathcal{E}_{M,s}[\Omega]$ as long as $R_1 \in \mathcal{E}_{M,s}[\Omega]$ and $R_2 \in \mathcal{E}_{M,s}[\Omega]$. If $R \in \mathcal{N}_s[\Omega]$ then obviously $DR \in \mathcal{N}_s[\Omega]$. $\mathcal{N}_s[\Omega]$ is a vector subspace of $\mathcal{E}_{M,s}[\Omega]$ and we have much more : if $R_1 \in \mathcal{E}_{M,s}[\Omega]$ and $R_2 \in \mathcal{N}_s[\Omega]$ their pointwise product $R_1 R_2$ is in $\mathcal{N}_s[\Omega]$ (in other words $\mathcal{N}_s[\Omega]$ is an ideal of the algebra $\mathcal{E}_{M,s}[\Omega]$).

We define the (simplified) generalized functions on Ω as the elements of the quotient algebra

$$\mathcal{G}_s(\Omega) = \frac{\mathcal{E}_{M,s}[\Omega]}{\mathcal{N}_s[\Omega]}.$$

Setting $R_\varepsilon(x) = R(\varepsilon,x)$ the representatives can be denoted by $\{R_\varepsilon\}_{0<\varepsilon<1}$ or R.

If $f \in \mathcal{C}^\infty(\Omega)$ then we consider the map $R \in \mathcal{E}_{M,s}[\Omega]$ defined by $R(\varepsilon,x) = f(x)$ and we obtain at once an inclusion of $\mathcal{C}^\infty(\Omega)$ into $\mathcal{G}_s(\Omega)$.

We say that $G_1, G_2 \in \mathcal{G}_s(\Omega)$ are associated (notation $G_1 \approx G_2$) if there are representatives R_1, R_2 of G_1, G_2 respectively such that for all $\psi \in \mathcal{D}(\Omega)$

$$\int_\Omega (R_1(\varepsilon,x) - R_2(\varepsilon,x))\, \psi(x)\, dx \to 0 \text{ when } \varepsilon \to 0$$

(then the same holds for all representatives of G_1 and G_2). Let T be a distribution on Ω. A generalized function $G \in \mathcal{G}_s(\Omega)$ is said to have T as macroscopic aspect iff for all $\psi \in \mathcal{D}(\Omega)$

$$\int R(\varepsilon,x)\, \psi(x)\, dx \to T(\psi) \text{ when } \varepsilon \to 0.$$

The inclusion $\mathcal{G}_s(\Omega) \subset \mathcal{G}(\Omega)$ is obtained as follows : to any $\varphi \in \mathcal{A}_0$ one can uniquely associate $\chi \in \mathcal{A}_0$ and $\varepsilon > 0$ such that $\varphi = \chi_\varepsilon$ (i.e. $\varphi(x) = \frac{1}{\varepsilon^n} \chi(\frac{x}{\varepsilon})$) and $\int |\chi(x)|^2\, dx = 1$ (or other similar properties that give uniqueness of χ and ε such that $\varphi = \chi_\varepsilon$). Then if $G \in \mathcal{G}_s(\Omega)$, if R is a representative of G, set $\overline{R}(\varphi,x) = R(\varepsilon,x)$; then to G associate the class $\overline{G} \in \mathcal{G}(\Omega)$ of $\overline{R}$. Since the map $G \to \overline{G}$ is injective one can identify $\mathcal{G}_s(\Omega)$ as a subalgebra of $\mathcal{G}(\Omega)$. For details see Biagioni [1] §1.8.

Remark. An even simpler formulation is in Egorov [1] : $\mathcal{N}_s[\Omega]$ there consists only of the maps R for which $R(\varepsilon,x) = 0$ for $\varepsilon > 0$ small enough. This permits an important simplification in definitions at the price of a loss of certain more evolved properties still contained in the $\mathcal{G}_s$ case described here (see for instance one of then in Colombeau-Méril [2]. The simplified formulation in Egorov [1] looks close to Nonstandard Analysis.

Research Problems. Various improvements in the definition of $\mathcal{G}(\Omega)$ are possible ; several of them are presented in Biagioni [1]. Also the sets $\mathcal{A}_q$ are not invariant under nonlinear (C^∞) changes of coordinates ; this can be easily solved by defining them with inequalities of the kind $|\int x^i \varphi(x)\, dx| \leq$ cste ε^q if $|i| \leq q$. In short the "best" definition of $\mathcal{G}(\Omega)$, if any, has not yet been clarified. Interesting works could be done in this direction.

§ 8.5 ANOTHER INTRODUCTION TO THESE GENERALIZED FUNCTIONS

The aim of this section is to provide an introduction to these generalized functions which is more elementary than the one in § 8.1. To open smoothly a way for a concept of generalized functions that could be freely multiplied (as well as differentiated) let us present the following way of viewing the distributions. Let

$$S = \{ \text{ sequences } (f_\varepsilon)_{0<\varepsilon<1} \text{ of } \mathscr{C}^\infty \text{ functions on } \mathbb{R} \text{ such that } \forall \varphi \in \mathfrak{D}$$

$$\int_{-\infty}^{+\infty} f_\varepsilon(x)\, \varphi(x)\, dx \text{ converges when } \varepsilon \to 0 \text{ to a number } T(\varphi) \}.$$

One can actually prove that T so defined is a distribution. Let

$$S_0 = \{ \text{sequences } (f_\varepsilon)_{0<\varepsilon<1} \in S \text{ such that } \forall \ \varphi \in \mathfrak{D} \quad \int_{-\infty}^{+\infty} f_\varepsilon(x)\, \varphi(x)\, dx \to 0 \text{ when } \varepsilon \to 0 \}.$$

S and S_0 are vector spaces for the componentwise addition and scalar multiplication. In S we define an equivalence relation R by (f_ε) R (g_ε) iff $(f_\varepsilon - g_\varepsilon) \in S_0$. The set of all equivalence classes is a vector space denoted by $\frac{S}{S_0}$. One can prove that this quotient space $\frac{S}{S_0}$ is isomorphic to the space $\mathfrak{D}'$ (indeed on proves that for any distribution T there is a sequence (f_ε) of $\mathscr{C}^\infty$ functions on $\mathbb{R}$ such that

$$\forall \ \varphi \in \mathfrak{D} \quad \int_{-\infty}^{+\infty} f_\varepsilon(x)\, \varphi(x)\, dx \to T(\varphi) \text{ when } \varepsilon \to 0).$$

How to modify S and S_0 so that the quotient space would be an algebra (i.e. a set with two internal laws + and · satisfying the usual properties), and even a differential algebra (i.e. an algebra with also an internal derivation operator satisfying the usual properties : the set $\mathscr{C}^\infty$ of all $\mathscr{C}^\infty$ functions on $\mathbb{R}$ is a differential algebra) ? For this, one requires the large set $\mathscr{L}$ (analogous to S) to be an algebra (S is not) and the small set $\mathscr{L}_0$ (analogous to S_0) to have the property that $a.b \in \mathscr{L}_0$ as soon as $a \in \mathscr{L}, b \in \mathscr{L}_0$ (i.e. $\mathscr{L} . \mathscr{L}_0 \subset \mathscr{L}_0$; one says in this case that $\mathscr{L}_0$ is an ideal of $\mathscr{L}$; this expresses that the equivalence relation R (modulo $\mathscr{L}_0$) is compatible with the multiplication in $\mathscr{L}$).

The simplest construction is as follows Laugwitz [1], Laugwitz - Schmielden [1] , Egorov [1]. Let

$$\mathscr{E} = \{ (f_\varepsilon)_{0<\varepsilon<1} \text{ where } f_\varepsilon \text{ are } \mathscr{C}^\infty \text{ functions} \}$$

and let

$$\mathscr{I} = \{ (f_\varepsilon)_{0<\varepsilon<1} \text{ such that } f_\varepsilon = 0 \text{ for } \varepsilon > 0 \text{ small enough } \}.$$

$\mathscr{I}$ is an ideal of $\mathscr{E}$ and so the quotient space $\frac{\mathscr{E}}{\mathscr{I}}$ is an algebra ; further it is a differential algebra by differentiating componentwise. To a $\mathscr{C}^\infty$ function f one can associate the constant sequence $f_\varepsilon = f$: the differential algebra $\mathscr{C}^{\infty}$ is a differential subalgebra of $\frac{\mathscr{E}}{\mathscr{J}}$. Let us choose a function $\rho \in \mathscr{D}$, of integral

one ($\int\limits_{-\infty}^{+\infty} \rho(x)\,dx = 1$) ; we set $\rho_{\varepsilon}(x) = \frac{1}{\varepsilon}\rho(\frac{x}{\varepsilon})$.   Let f be a continuous function on $\mathbb{R}$; then we set

$f_{\varepsilon} = f * \rho_{\varepsilon}$, the convolution product of f and ρ_{ε}, defined by $(f * \rho_{\varepsilon})(x) = \int\limits_{-\infty}^{+\infty} f(x-y)\,\rho_{\varepsilon}(y)\,dy$. It is

easy to prove that f_{ε} is a $\mathscr{C}^{\infty}$ function. To the continuous function f we associate the sequence

$(f_{\varepsilon})_{0<\varepsilon<1} \in \mathscr{E}$. This gives an inclusion of $\mathscr{C}$ into $\frac{\mathscr{E}}{\mathscr{J}}$. If T is a distribution then the convolution $f_{\varepsilon} = T$

$* \rho_{\varepsilon}$ is defined similarly and is also a $\mathscr{C}^{\infty}$ function. This gives an inclusion $\mathscr{D}' \subset \frac{\mathscr{E}}{\mathscr{J}}$. A defect of this

construction lies in that the direct inclusion of $\mathscr{C}^{\infty}$ into $\frac{\mathscr{E}}{\mathscr{J}}$ and the inclusion as a subspace of $\mathscr{C}$ do not

give the same result : $f * \rho_{\varepsilon} \neq f$ in general whatever the smallness of ε. Further the choice of the

function ρ (called a "mollifier") is arbitrary. Historically Laugwitz's construction is considered as a

precursor of Nonstandard Analysis, which then diverged from the purpose of this book.

A more refined construction permits to make the two above inclusions coherent. We set

$$\mathscr{E}_M = \{(f_{\varepsilon})_{0<\varepsilon<1} \in \mathscr{E} \text{ such that for each finite interval I, } \forall n \in \mathbb{N} \quad \exists N \in \mathbb{N},$$

$$\eta > 0 \text{ and } c > 0 \text{ such that } \sup_{x \in I} |f_{\varepsilon}^{(n)}(x)| \leq \frac{c}{\varepsilon^N} \text{ if } 0<\varepsilon<\eta \}.$$

The subscript M in $\mathscr{E}_M$ stands for "moderate" due to the (moderate) growth in powers of the variable

$\frac{1}{\varepsilon}$ when $\varepsilon \to 0$. $\mathscr{E}_M$ is a differential algebra for componentwise operations. We define an ideal $\mathscr{N}$ of

$\mathscr{E}_M$ by :

$$\mathscr{N} = \{(f_{\varepsilon})_{0<\varepsilon<1} \text{ such that } \forall I, \forall n, \forall q \in \mathbb{N} \quad \exists c>0, \eta>0 \text{ such that}$$

$$\sup_{x \in I} |f_{\varepsilon}^{(n)}(x)| \leq c\,\varepsilon^q \text{ if } 0<\varepsilon<\eta\}$$

i.e. the objects in $\mathscr{N}$ have a fast decay (faster than any power of ε) when $\varepsilon \to 0$. The quotient $\mathscr{G} = \frac{\mathscr{E}_M}{\mathscr{N}}$

is a differential algebra. $\mathscr{C}^{\infty}$ is contained in $\mathscr{G}$ from the constant sequence $f_{\varepsilon} = f$. $\mathscr{C}$ (and also $\mathscr{D}'$) is contained in $\mathcal{G}$ from the sequence $f_\varepsilon = f * \rho_\varepsilon$. The novelty relatively to $\frac{\mathcal{E}}{\mathcal{I}}$ is that a suitable choice of

the function ρ ensures coherence of the two inclusions of $\mathcal{C}^\infty$ into $\mathcal{G}$: since

$$\int_{-\infty}^{+\infty}\rho(\mu)d\mu=1, \quad (f * \rho_\varepsilon)(x) - f(x) = \int_{-\infty}^{+\infty}(f(x+\varepsilon\mu) - f(x))\,\rho(\mu)\,d\mu \;;$$

if $\int_{-\infty}^{+\infty}\mu^i\,\rho(\mu)\,d\mu = 0$ $\forall i \geq 1$ then Taylor's formula applied to f ensures that $f * \rho_\varepsilon - f \in \mathcal{N}$. A

minor fact should be mentioned : there does not exist a $\rho \in \mathcal{D}$ with the above properties (hint of

proof : if $\hat{\rho}$ is the Fourier transform of ρ these requirements can be stated as

$\hat{\rho}(0) = 1$, $\hat{\rho}^{(i)}(0)=0$ $\forall i \geq 1$; if $\rho \in \mathcal{D}$ $\hat{\rho}$ is analytic hence identical to 1 from the uniqueness

of analytic continuation). This fact is easily repaired by choosing ρ in a space of functions that have a

fast decay at infinity or by allowing various $\rho \in \mathcal{D}$: each of them satisfies requirements $\int_{-\infty}^{+\infty} \mu^i\,\rho(\mu)$

$d\mu = 0$ for $i \leq q$, for finite but arbitrarily large q depending on ρ. Taking all such ρ then the

functions $x \to f_\varepsilon(x)$ (in $\mathcal{E}_M$ and $\mathcal{N}$) have to be replaced by a set of functions $x \to f_\varepsilon(x,\rho)$ depending

on ρ ; further the arbitrariness in a choice of ρ disappears since all possible ρ's are considered. Note

that the original ideas leading to $\mathcal{G}$ from distribution theory are different from this presentation (see

§8.1). A dozen of variants of this space $\mathcal{G}$ have been introduced. One has a chain of inclusions $\mathcal{C}^\infty \subset$

$\mathcal{C} \subset \mathcal{D}' \subset \mathcal{G}$. $\mathcal{G}$ induces on $\mathcal{D}'$ the derivatives in the sense of distributions and $\mathcal{C}^\infty$ is a faithful

subalgebra of $\mathcal{G}$. $\mathcal{C}$ is not a subalgebra of $\mathcal{G}$: this is a more sophisticated version of the argument

presented at the beginning of this paper according to which the equality $Y^n = Y$ leads to a

contradiction , see §3.1. If f, g are two continuous functions, their classical product in $\mathcal{C}$ and their

new product in $\mathcal{G}$ are in general different elements of $\mathcal{G}$. But they are related in the so called

"association sense" defined as follows : two generalized functions F, G $\in$ $\mathcal{G}$, of respective

representatives $(f_\varepsilon)_{0<\varepsilon<1}$, $(g_\varepsilon)_{0<\varepsilon<1}$ are said to be associated iff

$$\forall \psi \in \mathcal{D} \int (f_\varepsilon(x) - g_\varepsilon(x))\,\psi(x)\,dx \to 0 \text{ as } \varepsilon \to 0.$$

We write $F \approx G$.

Example $Y^n \approx Y$ if $n \geq 1$ (and $Y^n \neq Y$ if $n \neq 1$ from the calculation above).

Setting $S_0' = \{(f_\varepsilon)_{0<\varepsilon<1} \in \mathcal{E}_M$ such that $\forall \, \psi \in \mathfrak{D} \int f_\varepsilon(x) \, \psi(x) \, dx \to 0$ when $\varepsilon \to 0 \}$ then by definition $F \approx G$ iff $(f_\varepsilon - g_\varepsilon)_{0<\varepsilon<1} \in S_0'$. S_0' is far larger than $\mathcal{N}$ and it is not an ideal of $\mathcal{E}_M$. The quotient space $\dfrac{\mathcal{E}_M}{S_0'}$ looks like $\mathfrak{D}'$: it is only a vector space (with a differentiation) that contains $\mathfrak{D}'$; its elements are classes of generalized functions ($\in \mathcal{G}$) that are associated with each other. To have a differential algebra one needs the more refined quotient $\mathcal{G} = \dfrac{\mathcal{E}_M}{\mathcal{N}}$. The unavoidable price to pay is that, since the objects in $\mathcal{G}$ are more refined (than the equivalence classes modulo S_0') there appears

a difference between the classical products and the new ones in $\mathcal{G}$ ($Y^n \neq Y$ in $\mathcal{G}$). A deeper analysis in various situations in which classical products exist reveals that the new products are associated to the classical ones, see Oberguggenberger [1] for instance.

<u>Exercise 1.</u> Consider the equation (E) $\dfrac{\partial u}{\partial t} + u \dfrac{\partial u}{\partial x} = 0$ (with various senses of the equality symbol) and seek a "travelling wave" solution of the form $u(x,t) = (u_r - u_l) \, Y \, (x - ct) + u_l$ (u_r, u_l, c real numbers) :

(a) Interpreting equation (E) as $\dfrac{\partial}{\partial t} u + \dfrac{\partial}{\partial x} \left(\dfrac{1}{2} u^2 \right) = 0$ in the sense of distributions $\left(\text{i.e. } \forall \, \psi \in \mathfrak{D} \right.$

on $\mathbb{R}^2$, $\int \left(u(x,t) \dfrac{\partial}{\partial t} \psi \, (x,t) + \dfrac{1}{2} u^2(x,t) \dfrac{\partial}{\partial x} \psi \, (x,t) \right) dx \, dt = 0 \big)$ prove that u is solution iff $c = \dfrac{u_r + u_l}{2}$.

(b) multiply (E) by u and then interpret it as $\dfrac{\partial}{\partial t} \left(\dfrac{1}{2} u^2 \right) + \dfrac{\partial}{\partial x} \left(\dfrac{1}{3} u^3 \right) = 0$ in the sense of distributions ; does one obtain the same formula, if $u_r \neq u_l$?

(c) Now interpret (E) with the equality in $\mathcal{G}$ (relative to $\mathbb{R}^2$: an immediate extension of the one-dimensional case considered here). Deduce from the same calculations as in (a) (b) that it has no travelling wave solution if $u_r \neq u_l$.

(d) Interpret (E) with the association in $\mathcal{G}$ ($\mathbb{R}^2$) i.e. $\dfrac{\partial u}{\partial t} + u \dfrac{\partial u}{\partial x} \approx 0$. Show that it has the above travelling wave solution iff $c = \dfrac{u_r + u_l}{2}$. Compute the travelling wave solutions of $u u_t + u^2 u_x \approx 0$ (one obtains the same formula as in (b) ; there is no contradition since $u_t + u u_x \approx 0$ does not imply $u u_t + u^2 u_x \approx 0$: incompatibility between the association and the multiplication).

<u>Exercise 2.</u> Consider the system
$$\begin{cases} \rho_t + (\rho u)_x & \approx 0 \\ (\rho u)_t + (\rho u^2)_x & \approx \sigma_x \\ \sigma_t + u \sigma_x & \approx u_x \end{cases}$$
which can be considered as a simplified model of elasticity. The purpose of this exercise is to seek travelling wave solutions, i.e. solutions ρ, u, σ which are constant on both sides of a discontinuity propagating at constant speed c. Such solutions are meaningless within the context of distributions

because of the product $u\,\sigma_x$ (one can be easily convinced of this fact : indeed this term can be interpreted as a product of the kind Y^n. Y' for arbitrary n since u can be represented in the form $u(x,t) = \Delta u\,Y^n\,(x-ct) + u_1$, $\Delta u, u_1 \in \mathbb{R}$, Y the Heaviside function, and various jump conditions are obtained for various values of n). In $\mathcal{G}(\mathbb{R}^2)$ travelling wave solutions are obtained by setting

$$\begin{cases} \rho(x,t) = \Delta\rho\ H(x-ct) + \rho_1 \\ u(x,t) = \Delta u\ K(x-ct) + u_1 \\ \sigma(x,t) = \Delta\sigma\ L(x-ct) + \sigma_1 \end{cases}$$

with $\Delta\rho, \Delta u, \Delta\sigma, \rho_1, u_1, \sigma_1, c \in \mathbb{R}$ and where H, K, L $\in \mathcal{G}(\mathbb{R})$ share the following properties :

- their restriction to $(-\infty, 0[$ is identical to 0 (the definition of the restriction of an element of $\mathcal{G}(\mathbb{R})$ to any open interval is quite obvious).
- their restriction to $]0,+\infty)$ is identical to 1
- they admit representatives H^ε (K^ε, L^ε respectively) such that $\displaystyle\sup_{\substack{0<\varepsilon\leq1 \\ x\in\mathbb{R}}} |H^\varepsilon(x)| < +\infty$ (so as to eliminate a singularity such as a Dirac delta function at the origin).

Those H, K, L satisfying these properties are called "Heaviside generalized functions" ; example : for every $n = 1,..$ Y^n is such a function if Y is the Heaviside function. It is clear that it would be abusive to state at first $H = K = L$ from the sole assumption that one seeks travelling wave solutions.

<u>Question (a)</u>. Study the solutions of the above type ; show that the jump conditions depend on an arbitrary real parameter A defined by $HK' \approx A\delta$ (δ the Dirac distribution).

<u>Question (b)</u>. State the system in the form

$$\begin{cases} \rho_t + (\rho u)_x = 0 \\ (\rho u)_t + (\rho u^2)_x = \sigma_x \\ \sigma_t + u\sigma_x \approx u_x \end{cases}$$

(which is justified physically in §4.3); compute the number A in that case. Solutions are given in §4.4.

A number of such exercises, of physical significance, are given in §3.5 and §4.7. Although inaccessible within distribution theory, the study of their travelling wave solutions is amazingly simple, and rich of information for numerical solutions of systems of engineering.

REFERENCES

M. Adamczewski [1] - Vectorisation, analyse et optimisation d'un code bidimensionnel eulérien. Thèse, Bordeaux, 1986.

M. Adamczewski- J. F. Colombeau - A. Y. Le Roux [1] - Convergence of numerical schemes involving powers of the Dirac delta function. J. Math. Anal. Appl. 145, 1, 1990, p 172-185.

J. Aragona [1] - Théorème d'existence pour l'opérateur $\bar{\partial}$ sur les formes différentielles généralisées. C. R. Acad. Sci. Paris Sér. I, Math, 300 (1985) p. 239 - 242.

J. Aragona [2] - On existence theorems for the $\bar{\partial}$ operator on generalized differential forms. Proc. London Math. Soc. (3) 53 (1986) p. 474-488.

J. Aragona - H. A. Biagioni[1] - An intrinsic definition of the Colombeau algebra of generalized functions. Analysis Mathematica 17, 1991, p75-132.

J. Aragona - J. F. Colombeau [1] - On the $\bar{\partial}$ Neumann problem for generalized functions J. Math. Anal. Appl. 110, 0, 1985, p. 179-199.

J. Aragona - J. F. Colombeau [2] - The interpolation theorem for holomorphic generalized functions. Anales Polonici Mathematici XLIX (1988) p. 151-156.

J. Aragona - F. Villareal [1] - Colombeau's theory and shock waves in a problem of hydrodynamics. J. Anal. Math. in press.

L. Arnaud [1] Quelques schémas numériques adaptés à l'élastodynamique en configuration axisymétrique, Thèse, Bordeaux, 1990

F. Bampi, C. Zordan [1] Higher order shock waves, preprint

Y.A. Barka [1] - Thèse, Bordeaux 1988.

Y. A. Barka- J. F. Colombeau - B. Perrot [1] - A numerical modelling of the fluid/fluid acoustic dioptra. Journal d'Acoustique. 2, 1989, p 333-346.

R. Baraille, G. Bourdin, F. Dubois, A.Y. Le Roux [1] - Une version à pas fractionnaires du schéma de Godunov pour l'hydrodynamique. Comptes Rendus Acad. Sci. Paris, t 314, 1, 1992, p 147-152.

D. Bedeaux, A. M. Albano, P. Mazur [1] Boundary conditions and non equilibrium thermodynamics. Physica A, 82, 1978, p 438-462.

F. Berger [1]. Thèse Lyon-St Etienne 1993.

H. A. Biagioni [1] A Nonlinear Theory of Generalized Functions. Lecture Notes in Mathematics 1421, Springer Verlag, Berlin - Heidelberg - New York 1990

H. A. Biagioni [2] - The Cauchy problem for semilinear hyperbolic systems with generalized functions as initial conditions. Resultate Math. 14 (1988) p. 231-241.

H. A. Biagioni - J. F. Colombeau [1] - Borel's theorem for generalized functions. Studia Math. 81 (1985) p. 179-183.

H. A. Biagioni - J. F. Colombeau [2] - Whitney's extension theorem for generalized functions. J. Math. Anal. Appl. 114,2 (1986) p. 574-583.

H. A. Biagioni - J. F. Colombeau [3] - New generalized functions and C^{∞} functions with values in generalized complex numbers. J. London Math. Soc (2) 33 (1986) p. 169-179.

H. A. Biagioni - R.J. Iorio [1]. Generalized solutions of the Benjamin - Ono and Smith equations.

H. A. Biagioni - M. Oberguggenberger [1] Generalized solutions to Burgers' equation, J. Diff. Eq. in press.

H. A. Biagioni - M. Oberguggenberger [2] Generalized solutions to the Korteweg de Vries and the regularized long wave equations, preprint

E. Bonnetier [1] Formulation of the equations for elastic flows. Private Communication May 1989.

N. N.Bogoliubov. D. V. Shirkov [1] Introduction to the Theory of Quantized Fields, Interscience, several editions

I. Bouvier [1] Un schéma tridimensionnel adapté à la propagation d'ondes élastiques. Thèse, Bordeaux, 1988

J. J. Cauret [1] Analyse et développement d'un code bidimensionnel élastoplastique. Thèse, Bordeaux, 1986.

J. J. Cauret, J. F. Colombeau, A. Y. Le Roux [1] Discontinuous generalized solutions of nonlinear nonconservative hyperbolic equations. J. Math. Anal. Appl. 139, 2, 1989 p 552-573.

J. F. Colombeau [1] - Differential Calculus and Holomorphy. Real and Complex Analysis in Locally Convex Spaces. North-Holland Math. Studies 64, 1982.

J. F. Colombeau [2] - New Generalized Functions and Multiplication of Distributions. North-Holland Math. Studies 84, 1984.

J. F. Colombeau [3] - Elementary Introduction to New Generalized Functions. North-Holland Math. Studies 113, 1985.

J. F. Colombeau [4] - New general existence results for partial differential equations (100 p.), unpublished paper, 1986.

J. F. Colombeau [5] - A multiplication of distributions. J. Math. Anal. Appl. 94, 1 (1983) p. 96-115.

J. F. Colombeau [6] - New generalized functions. Multiplication of distributions. Physical applications. Portugal. Math. 41, 1-4 (1982) p. 57-69.

J. F. Colombeau [7] - Une multiplication générale des distributions. C. R. Acad. Sci. Paris Sér. I, Math. 296 (1983) p. 357-360.

J. F. Colombeau [8] - Some aspects of infinite dimensional holomorphy in mathematical physics. In "Aspects of Mathematics and its Applications", editor J. A. Barroso, North-Holland Math. Library 34 (1986) p. 253-263.

J. F. Colombeau [9] - A new theory of generalized functions. In "Advances of Holomorphy and Approximation Theory", editor J. Mujica, North-Holland Math. Studies 125 (1986) p. 57-66.

J. F. Colombeau [10] - Nouvelles solutions d'équations aux dérivées partielles. C. R. Acad. Sci. Paris Sér. I, Math. 301 (1985) p. 281-283.

J. F. Colombeau [11] - Multiplication de distributions et acoustique. J. d'Acoustique 1-2 (1988) p. 9-14.

J. F. Colombeau [12] - Generalized functions, multiplication of distributions, applications to elasticity, elastoplasticity, fluid dynamics and acoustics. In : "Generalized Functions, Convergence Structures and their Applications". B. Stankovic, E. Pap, S. Pilipovic, V. S. Vladimirov (editors), Plenum Press, New York (1988) p. 13-28.

J. F. Colombeau [13] - The elastoplastic shock problem as an example of the resolution of ambiguities in the multiplication of distributions. J. Math. Phys. 30, 10, 1989, p 2273-2279.

J. F. Colombeau [14] - Multiplication of Distributions. Bull. A.M.S. 23, 2, 1990, p. 251-268.

J. F. Colombeau [15] - Introduction to "new generalized functions and multiplications of distributions . World Scientific Pub. Com., Singapore, ICPAM Lecture Notes (1988) p. 338-380.

J. F. Colombeau - J. E. Galé [1] - Holomorphic generalized functions. J. Math. Anal. Appl. 103, 1 (1984) p. 117-133.

J. F. Colombeau - J. E. Galé [2] - The analytic continuation for generalized holomorphic functions. Acta Math. Hung. 52, 1-2 (1988) p. 57-60.

J. F. Colombeau - A. Heibig [1] - Nonconservative products in bounded variation functions. SIAM J. of Math. Anal.

J.F. Colombeau - A. Heibig - M. Oberguggenberger [1] - Generalized solutions to partial differential equations of evolution type.

J. F. Colombeau - M. Langlais [1] - Existence et unicité de solutions d'équations paraboliques nonlinéaires avec conditions initiales distributions. C. R. Acad. Sci. Paris Sér. I Math. 302 (1986) p. 379-382.

J. F.Colombeau. M. Langlais [2] - Generalized solutions of nonlinear parabolic equations with distributions as initial conditions. J. Math. Anal. Appl. 145, 1, 1990, p 186-196.

J. F. Colombeau - J. Laurens- B. Perrot [1] - Une méthode numérique de résolution des équations de l'acoustique dans un milieu à caractéristiques $\mathcal{C}^\infty$ par morceaux. Colloque de Physique, supplément au J. de Physique 2, tome 51, 1990, p 1227-1230.

J. F. Colombeau - A. Y. Le Roux [1] - Numerical techniques in elastodynamics. Lecture Notes in Math. 1270, Springer (1987) p. 103 - 114.

J. F. Colombeau - A. Y. Le Roux [2] - Numerical methods for hyperbolic systems in nonconservative form using products of distributions. Advances in Computer Methods for Partial Differential Equations VI, IMACS (1987), R. Vichnevetsky and R. S. Stepleman (editors), p. 28 - 37.

J. F. Colombeau - A. Y. Le Roux [3] - Multiplications of distributions in Elasticity and Hydrodynamics. J. Math. Phys. 29, 2 (1988) p. 315 - 319.

J. F. Colombeau - A. Y. Le Roux - B. Perrot [1] - Multiplications de distributions et ondes de choc élastiques ou hydrodynamiques en dimension un. C. R. Acad. Sci. Paris Sér. I. Math. 305 (1987) p. 453 - 456.

J. F. Colombeau - A. Y. Le Roux - A. Noussair - B. Perrot [1] - Microscopic profiles of shock waves and ambiguities in multiplications of distributions. SIAM J. Num. Anal. 26,4 (1989) p. 871 - 883.

J.F. Colombeau - A. Méril [1] - Generalized functions on C^∞ manifolds.

J.F. Colombeau - A. Méril [2] - Generalized functions solutions of linear PDEs with analytic coefficients.

J. F. Colombeau - M. Oberguggenberger [1] - Generalized functions and products of distributions. preprint.

J. F. Colombeau - M. Oberguggenberger [2] - Hyperbolic systems with a compatible quadratic term, delta waves and multiplication of distributions. Comm. in PDEs 15, 7, 1990, p. 905-938.

J. F. Colombeau - M. Oberguggenberger [3] - Generalized solutions and measure valued solutions to conservation laws. preprint.

G. Dal Maso - Ph. Le Floch - F. Murat [1] - Definition and weak stability of a nonconservative product. preprint.

M. De Billy [1]. Private communications on experimental results in acoustics diffusion.

M. Defourneaux [1] - Théorie et mesure des ondes de choc dans les solides. ENSTA, Paris, 1975, cours CP 32.

P. De Luca [1] - Modélisation numérique en élastoplasticité dynamique. Thèse, Bordeaux, 1989.

J. C. Diniz Fernandes [1] - O teorema de extensao de Hartogs para funçoes holomorfas generalizadas. Thesis, Sao Paulo, 1985.

R. Di Perna [1] - Measure valued solutions to conservation laws. Arch. Rat. Mech. Anal. 88,3 (1985) p. 223 - 270.

R. Di Perna - A. Majda. [1] Oscillations and concentrations in weak solutions of the incompressible fluid equations. Comm. Math. Phys 108, 1987, p 667 - 689

P. A. M. Dirac [1] The physical interpretation of the quantum dynamics. Proc. Royal. Soc. London, section A, 113, 1926 - 27, p. 621 - 641.

P. A. M. Dirac [2] Theory of emission and absorption of radiation. Proceed. Royal Soc. London, série A, 114, 1927, p 221 - 295.

Y. Egorov [1] A theory of generalized functions. Uspehi Math. Nauk (russian) 45, 5, 1990, p. 3-40. Russian Math. Surveys in press.

Ph. Le Floch [1] Entropy weak solutions to nonlinear hyperbolic systems in nonconservation form.Comm. in P. D. E. s 13 (6) 1988 p 669-722

Ph. Le Floch [2] : Entropy weak solutions to nonlinear hyperbolic systems in nonconservation form. Nonlinear Hyperbolic Equations. . . J. Ballman, R. Jeltsch editors. Notes on Numerical Fluid Mechanics Vol. 24, Vieweg 1989.

Ph. Le Floch [3] : Shock waves for nonlinear hyperbolic systems in nonconservation form. SIAM J. of Applied Mathematics, in press.

Duret - D. A. Matuska [1] Finite difference solutions to the equations of continuum mechanics A. F. A. T. L. Trans. 42 - 125

T. Gramchev [1] Nonlinear maps in spaces of distributions. Math. Zeit. in press.

K. H. Hain [1] The partial Donor cell method. J. Comp. Phys. 73, 1987, p 131 - 147.

Heaviside [1] On operators in mathematical physics. Proc. Royal. Soc. London, 52, 1893, p 504 - 529, znf 54, 1894, p 105 - 143.

L. Hörmander [1]. An Introduction to Complex Analysis in Several Varaiables, North Holland 1973

L. Hörmander [2]. The Analysis of Linear Partial Differential Operators Vol I, II, III, Springer 1983.

A. E. Hurd - P. A. Loeb [1] An Introduction to Nonstrandard Real Analysis. Academic Press, 1985.

J. Jelinek [1] - Characterization of the Colombeau product of distributions. Commentat. Math. Univ. Carol. 27,2 (1986) p. 377 - 394

J. Jelinek [2] - Distinguishing example for the Tillmann product of distributions. Comment. Math. Univ. Carolinae. 31 ,4, 1990, p. 693-700.

R. Jost [1] The General Theory of Quantized Fields. A. M. S. Lectures in Applied Math., IV, 1965, Providence, R. I.

D. Kastler [1] Introduction à l'Electrodynamique Quantique. Dunod, Paris, 1961.

B. L. Keyfitz-H. C. Kranzer [1] : A viscosity approximation to a system of conservation laws with no classical Riemann solution. In "Nonlinear hyperbolic problems" C. Carrasso ...editors. Lecture Notes in Mathematics 1402, Springer Verlag 1988, p 185-197.

C. O. Kiselman [1] - Sur la définition de l'opérateur de Monge-Ampère complexe. Lecture Notes in Math. 1094, Springer (1984) p. 139 - 150.

F. Lafon [1] - Generalized solution of a one dimensional semilinear hyperbolic system with discontinuous coefficients. Application to an electron transport problem. preprint.

F. Lafon - M. Oberguggenberger [1] - Generalized solutions to symmetric hyperbolic systems with discontinuous coefficients : the multidimensional case. J. Math. Anal. Appl.160, 1, 1991, p 93 - 106.

M. Langlais [1] - Generalized functions solutions of monotone and semilinear parabolic equations. Mona. für Math. in press.

D. Laugwitz [1] - Applications of infinitesimal numbers (german). J. Reine Angew. Math. 207, 1961, p56-60.

D. Laugwitz - C. Schmielden [1] - An extension of the infinitesimal calculus (german). Math. Zeitschr. 69, 1958, p1-39.

J. Laurens [1] Une modelisation numerique du dioptre acoustique liquide/solide. Colloque de physique, supplément au J. de Physique 2, tome 51, 1990, p1219-1222.

J. Laurens [2]. Private communication.

A. Y. Le Roux, P. De Luca [1] A velocity - pressure model for elastodynamics. In "Nonlinear Hyperbolic Equations, Theory, Computation Methods and Applications". J. Ballman and R. Jeltch editors. Notes on Numerical Fluid Mechanics vol 24, Vieweg Verlag 1989, p 374 - 383.

H. Lewy [1] An example of a smooth linear PDE without solution. Annals of Math. 66, 1, 1957, pp 155-158.

D. R. Liles-W. H. Reed. [1] - A semi implicit method for two phase fluid dynamics. J. Comp. Phys. 26, 1978, p390-407.

J. L. Lions - E. Magenes [1] Problèmes aux limites non homogènes, Dunod, Paris, 1970.

T. P. Liu, Ph. Le Floch [1] : Convergence of the random choice method for systems in nonconservation form, preprint.

J. J. Lodder [1] A simple model for a symmetrical theory of generalized functions V. Physica A, 11, 1982, p 404-410

A. Martineau [1] Sur les fonctionnelles analytiques et la transformation de Fourier Borel. J. Anal. Math. Jerusalem 11, 1963, p 1 - 164

A. Martineau [2] Les hyperfonctions de Pr. Sato. Séminaire Bourbaki 13è année, 1960 -61, n°214

D. A. Matuska [1] Hull user's manual, A. F. A. T. L. Trans. 84 (June 1984)

A. Marzouk [1] - Régularité de solutions généralisées d'équations différentielles algébriques, Thesis, Bordeaux, 1989.

A. Marzouk - B. Perrot [1]- Private communication, 1989.

A. Noussair [1] - Conception et validation de schémas numériques pour l'élastoplasticité dynamique en milieu non homogène. Thesis, Bordeaux, 1989.

M. Oberguggenberger [1] - Products of distributions. J. Reine Angew. Math. 365 (1986) p. 1-11.

M. Oberguggenberger [2] - Multiplication of distributions in the Colombeau algebra $\mathcal{G}(\Omega)$. Boll. Unione Mat. Ital. (6) 5-A (1986) p. 423-429.

M. Oberguggenberger [3] - Weak limits of solutions to semilinear hyperbolic systems. Math. Ann. 274 (1986) p. 5999-607

M. Oberguggenberger [4] - Generalized solutions to semilinear hyperbolic systems. Monatsh. Math. 103 (1987) p. 133-144

M. Oberguggenberger [5] - Solutions généralisées de systèmes hyperboliques semilinéaires. C. R. Acad. Sci. Paris Sér. I. Math. 305 (1987) p. 17. 18

M. Oberguggenberger [6] - Hyperbolic systems with discontinuous coefficients : generalized solutions and a transmission problem in acoustics. J. Math. Anal. Appl. 142 (1989) p. 452-467.

M. Oberguggenberger [7] - Hyperbolic systems with discontinuous coefficients : examples. In : " Generalized Functions, Convergence Structures and their Applications", B. Stankovic, E. Pap, S. Pilipovic, V. S. Vladimirov (editors). Plenum Press, New York, 1988, p. 257-266.

M. Oberguggenberger [8] - Systèmes hyperboliques à coefficients discontinus : solutions généralisées et une application à l'acoustique linéaire. C. R. Math. Acad. Sci. Canada, vol X, n°3 (1988) p . 143-148.

M. Oberguggenberger [9] - Products of distributions : nonstandard methods. Z. Anal. Anwendungen 7 (4) (1988) p. 347-365.

M. Oberguggenberger [10] The Carleman system with positive measures as initial data ; generalized solutions. Transport Th. Stat. Phys. in press.

M. Oberguggenberger [11] Multiplication of distributions and applications to partial differential equations. book. Innsbruck. 1991.

M.Oberguggenberger [12] Semilinear hyperbolic systems with rough initial data : generalized solutions. In "Generalized functions and convergence" edited by P. Antosik and A. Kaminski. World Scientific Publishers, Singapore 1990.

M.Oberguggenberger [13] Case study of a nonlinear, nonconservative, non strictly hyperbolic system. J. of Nonlinear Analysis T.M.A. in press.

M. Oberguggenberger - Y. G. Wang [1] Generalized solutions to scalar conservation laws and the gas dynamics system with viscosity terms.

M. Oberguggenberger - Y. G. Wang [2] Generalized solutions to the isentropic gas dynamics and elastodynamics systems.

B. J. Plohr - D. H. Sharp [1] A conservative eulerian formulation of the equations for elastic flow Preprint Los Alamos Nat. Lab 1988

B. Poirée [1] - Les équations de l'acoustique linéaire dans un fluide parfait au repos, à caractéristiques indéfiniment différentiables par morceaux. Rev. CETHEDEC 52 (1977) p. 69-79.

B. Poirée [2] - Représentations lagrangiennes et eulériennes de l'acoustique linéaire. Rev. CETHEDEC 56 (1978) p. 73-80.

B. Poirée [3] - Equations de perturbations et équations de passage associées. Application au dioptre acoustique. Rev. CETHEDEC 69 (1981) p. 1-49.

B. Poirée [4] - Les équations de l'acoustique linéaire et nonlinéaire dans les fluides en mouvement. Thèse es Sciences Physiques, Université de Paris VI, 1982.

C. K. Raju [1] Junction conditions in general relativity. J. Phys. A ; Math. Gen 15, 1982 ; p 1785 - 97

C. K. Raju [2] Products and composition with the Dirac delta function. J. Phys. A, Math Gen 1983 , p 3739-3753

J. Rauch - M. Reed [1] - Nonlinear superposition and absorption of delta waves in one space dimension. J. Funct. Anal. 73 (1987) p . 152 - 178.

R. D. Richtmyer [1] Principles of Advanced Mathematical Physics Vol 1. Springer Verlag. Berlin - Heidelberg New York 1978.

E. E. Rosinger [1] Generalized solutions of nonlinear partial differential equations. North Holland. Amsterdam - Oxford - New York 1987.

L. Sainsaulieu [1] private communication.

L. Schwartz [1] - Théorie des Distributions, Hermann, Paris, several editions.

L. Schwartz [2] Sur l'impossibilité de lamultiplication des distributions. Comptes Rendus Acad. Sci. Paris, 239, 1954, p 847-848.

D. Serre [1] Les ondes planes en electromagnérisme non linéaire, Physica D 31, 1988, p, 227-251.

P. Shapira [1] Une équation aux dérivées partielles sans solution dans l'espace des hyperfonctions. Comptes Rendus Acad. Sci. Paris 265 (1967) p 665-667.

S. Sobolev [1] Méthode nouvelle à résoudre le problème de Cauchy pour les équations hyperboliques normales. Math. Sbornik, 1, 1936, p 39-71.

G. A. Sod [1] A survey of several finite difference methods for systems of nonlinear hyperbolic conservation laws. J. Comput. Phys 27, 1978, p1-31.

H.B. Stewart - B. Wendroff. Two phase flow : models and methods. J. Comput. Phys. 56, 1984, p 363-409.

Streater - Whigtman [1] PCT, Spin, and statistics and all that Addison Wesley Pub. Comp. New York 1964.

T. D. Todorov [1] Colombeau's new generalized functions and Nonstandard Analysis. In "Generalized functions, convergence. Structures and applications". B. Stankovic, E. Pap, S. Pilipovic, V. S. Vladimirov editors) Plenum Pub. Comp, New York, 1988, p...

I. Toumi. Etude du problème de Riemann et construction de schémas numériques de type Godunov multidimensiojnels pour des modèles diphasiques. Thèse de Mathematiques Appliquées Ecole Polytechnique 1989

INDEX